MW00977137

Builder's Guide to Residential Plumbing

Other books in the Builder's Guide Series

Builder's Guide to Barriers: Doors, Windows, & Trim, by *Dan Ramsey*

Builder's Guide to Change-of-Use Properties by *R. Dodge Woodson*

Builder's Guide to Decks by *Leon Frechette*

Builder's Guide to Foundations & Floor Framing by *Dan Ramsey*

Builder's Guide to Residential Plumbing

R. Dodge Woodson

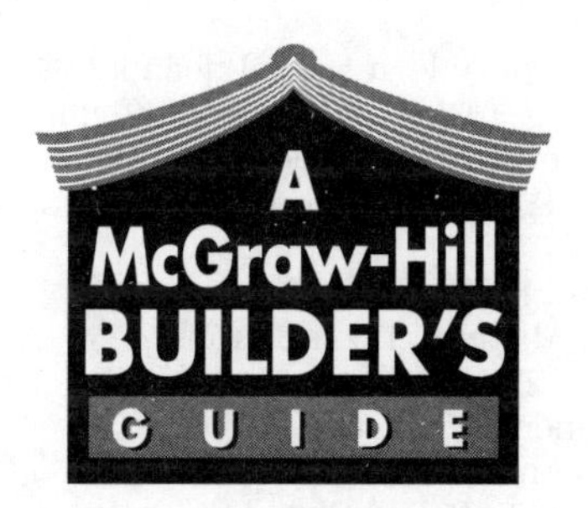

McGraw-Hill

New York San Francisco Washington, D.C. Auckland Bogotá
Caracas Lisbon London Madrid Mexico City Milan
Montreal New Delhi San Juan Singapore
Sydney Tokyo Toronto

McGraw-Hill

*A Division of The **McGraw-Hill** Companies*

Published by The McGraw-Hill Companies, Inc.

hc 1 2 3 4 5 6 7 8 9 0 DOC/DOC 9 0 0 9 8 7 6

Library of Congress Cataloging-in-Publication Data

Woodson, R. Dodge (Roger Dodge), 1955–
Builder's guide to residential plumbing / by R. Dodge Woodson.
p. cm.
Includes index.
ISBN 0-07-071781-8
1. Plumbing. 2. Building—Superintendence. I. Title.
TH6238.W659 1995
696'.1—dc20 95-25601
CIP

McGraw-Hill books are available at special quantity discounts to use as premiums and sales promotions, or for use in corporate training programs. For more information, please write to the Director of Special Sales, McGraw-Hill, 11 West 19th Street, New York, NY 10011. Or contact your local bookstore.

Acquisitions editor: April Nolan
Editorial team: Annette M. Testa, Book Editor
Susan W. Kagey, Supervising Editor
Lori Flaherty, Executive Editor
Production team: Katherine G. Brown, Director
Rhonda E. Baker, Coding
Toya B. Warner, Computer Artist
Rose McFarland, Desktop Operator
Nancy K. Mickley, Proofreading
Joann Woy, Indexer
Design team: Jaclyn J. Boone, Designer
Katherine Lukaszewicz, Associate Designer

0717818
GEN1

To Afton, Adam, and Kimberley, the most important people in the world to me. Kimberley provides help and support. Afton is very understanding and always gives me inspiration. Adam is a baby, but his smiles keep me going.

Acknowledgments

I would like to acknowledge and thank my parents, Maralou and Woody, for all the help they have given me over the years. Although they did not actually help me write this book, they did give me the opportunity to gain the knowledge and experience needed to make this project effective.

Contents

Appendices

Introduction

Plumbing is a vital element in many successful residential jobs. Whether you are building a new house, adding a room, or remodeling an attic or basement, plumbing will probably be involved. Many contractors are afraid of the plumbing aspects of their jobs because of bad experiences in the past or horror stories told to them by other contractors. Although some of these fears are justified, many aren't.

As a general contractor, you have a lot of responsibilities on your shoulders, and they can be a heavy burden. Some of the risks that you are exposed to in the plumbing end of a job can cause you to lose a lot of money, possibly even your whole business. Granted, most jobs go pretty well, and few general contractors go out of business because of the plumbing done on their jobs, but it can happen. Making a serious mistake can result in a lawsuit that might pull you under. Can you afford this? I doubt it. Fortunately, you are holding a tool in your hands right at this moment that can give you more control over your jobs.

Who am I? I'm a master plumber and general contractor. As a builder, I've constructed as many as 60 homes a year. My experience as a remodeling contractor is also very extensive. Plug in my plumbing experience, with more than 20 years in the field, and I'm pretty well qualified to help you. In addition to my field experience, I've taught plumbing code and apprentice classes at a technical college. When you wrap all of my credentials up into one package, they are extensive.

Do I know all there is to know about residential plumbing and building? Of course not. I don't think any one person can know all there is to know. Technology is changing so rapidly that a person has to stay actively involved to keep up-to-date. I'm still active in the trades, so I have a good feel for what's going on in the field.

How much do you know about plumbing? If you're an experienced general contractor, you have a better grasp of the trade than most people, but you probably don't know as much as your plumber.

You could, though, and with this knowledge and understanding about plumbing, you'll be less likely to suffer financial losses and problems on your jobs. That's what this book is all about. It is going to help you reduce lost time and money that are often attributed to jobs that don't run smoothly.

Is it possible to save time and money on your plumbing installations? Absolutely, and Chapter 1 shows you how. In this chapter, I show how you can save both time and money without giving up quality workmanship and materials.

The second chapter in this book offers a logical progression of events to follow in your next plumbing installation. This step is important in making more profits with a plumbing job. Because planning is paramount, as a general contractor, it is up to you to schedule plumbers effectively. With the help of this chapter, you can find out how to make the most of your jobs.

This book offers a wealth of information on a variety of topics such as coordinating subcontractors and employees efficiently; pinpointing various code-related issues; hiring your workers, whether they be employees, piece workers, or subcontractors; gearing up for and passing inspections; and adding plumbing to various parts of your house.

Each chapter deals with a different and specific aspect of the plumbing trade as it relates to you, the general contractor. The appendices provide tables and forms that you will find useful. Because I am both a plumber and general contractor, I can show you perspectives that other business advisors can't. I've seen the world from both sides of the table. This allows me to give you a better picture of what to expect in your business.

What will you get out of reading this book? It offers a lot of real-life experience and knowledge, which is not based on textbook theories but on my own personal experiences. There are technical facts about plumbing in the book, but you will learn more about how to make your business better and more profitable. If you want to make more money, this book can help you.

Take a few moments to study the topics listed in the table of contents. You will see that they all apply to the type of work with which residential contractors come in contact. By flipping through the pages, you will see that the book is written in a comfortable, conversational style with numerous illustrations and tables that help make the book even easier to read and understand. There is no cryptic code language in this book; instead, you get reader-friendly words.

If I had owned a book like this early in my career, I might not have made the many mistakes that gave me my first-hand experience. Avoiding costly mistakes could have allowed me to progress more rapidly in building a successful business. With the advice and data provided in this book, you have the opportunity to propel your business into a more desirable income level. Go ahead, read a few pages. I'm certain you will see the value of this book very quickly.

1

Saving time and money on plumbing installations

Wouldn't it be nice if you could save time and money on plumbing installations? If you've been in the contracting business for very long, I'm sure you've found that plumbers are expensive and independent workers. As a master plumber and plumbing contractor, I can attest to this fact. There are ways, however, for you, the general contractor, to take control and save both time and money on your next job.

Because I am both a master plumber and a general contractor who has done extensive remodeling and new construction plumbing in nearly 60 single-family homes each year, I am uniquely qualified to discuss plumbers and their requirements. This experience has allowed me to see plumbing from both sides of the table. For example, as a general contractor, I can see various plumbing situations from the client's perspective. Conversely, while looking at jobs as a plumber and plumbing contractor, I see situations from the plumber's perspective. As I combine my experience and perspective in this book, you wind up with the best of both worlds.

Because plumbing is a phase of work that many people can't readily grasp or aren't will to try, general contractors often end up paying plumbers a fair amount of money. There is no great mystery, though, surrounding the plumbing trade, at least not in technical terms. The trade is logical. If you are willing to invest a little time and effort in becoming familiar with plumbing principles, you can begin to reduce the many frustrations you may have toward pipe, water, and plumbers.

Plumbing requires a commitment to learn, much like anything worth doing. You don't, however, have to learn the trade inside and out or pass the test for your plumber's license to benefit from general knowledge about the trade. Many of the conflicts between plumbers and general contractors are caused by ignorance or negligence on the part of the contractors. If you learn to understand what plumbers must face on your jobs, you will be in a much better position to avoid problems and confrontations. Let me give you a quick example of what I'm talking about.

Let's say that you're a general contractor who is building a new house for a customer. You are responsible for the layout and construction. I'm your plumber. First of all, the two of us should get together before the house is started and discuss strategy. This type of meeting, however, is rare between contractors. It is more likely that the plumbers are not given a second thought until rough-ins are ready to be started. This is a mistake. To prove this, I'm going to give you two hypothetical situations that could cause a lot of trouble if plumbing routes are not mapped out before construction begins.

The first problem involves the first floor powder room. Because the space is limited in this room, there is not much choice when it comes to fixture placement. When I get on the job, there are no carpenters around and a problem arises. When laying out the plumbing, I find that a floor joist has been installed directly under the location where I must install a closet flange for the toilet. Because the toilet must have 15 inches of clear space from the center of the flange to either side wall and the room is only 30 inches wide, by code, I have no choice but to cut out the floor joist. So I cut it out and install my drain pipe and flange.

Now that I've cut out the joist, someone is going to have to fix the joist problem. Because your carpenters are piece workers, they are going to complain about having to do extra work in a tight crawl space that was not part of their original bid. Not only do they have to get headers put in, they will now have to work around my pipes. This makes their job even more difficult. What could you do, though? They weren't on the job when the joist had to be moved. After all, my contract addresses this issue clearly saying that all routes required for plumbing will be accessible for my installation.

As I'm roughing-in the house, your carpenters show up. I tell them about the joist I cut. At least I'm good enough to tell them. Some plumbers wouldn't have bothered and the framing inspection would have failed. After discussing the situation, the carpenters tell me that they're not going to fix the joist. I'm certainly not going to fix it, so you've got a problem. That's not the worst of the day, though.

As I proceed with my rough-ins in the house, I come to my second problem in an upstairs bathroom. This bathroom is situated over the main foyer of the home. Someone drew the plans so that the toilet is located over the middle of the foyer. Because the toilet needs a long horizontal drain that empties into a pipe that must then descend into the crawl space, a major problem arises. Had the contractor contacted me when drawing up these plans, I could have easily told the person to switch the location of the toilet with the adjacent lavatory so that the pipes could have gone down near the partition wall that comes up to form the foyer. But I was not consulted.

Due to the location of the toilet, I now have no way to get the drain from the vertical wall to the toilet without major carpentry work. The amount of pitch required on the pipe coming from the toilet drain will create a structural problem with the joists, and I will no longer be able to drill the joists. You know as well as I do that the top and bottom of the joists can't be left too thin without violating the building code. So, what am I to do?

If the lavatory and toilet locations are switched, I can make the plumbing work without major carpentry work. The smaller diameter of the pipe required for the lavatory will allow me to drill it through the ceiling joists without a problem. As the toilet is moved closer to the vertical wall, I gain enough distance to eliminate my problem with its drain. If I don't make this change, you are going to have to drop the ceiling in the foyer or use some steel to beef up the joists I drill. Under the circumstances, I'm at a standstill until we reach a decision on this issue. This causes me to lose money because I can't maintain my production schedule. Now I'm mad.

Look at what has happened here so far. Your carpenters are hostile toward me and my workers, and, once you are made aware of the problems, you will not be happy either. To top it all off, I'm mad because my crews are sitting around with nothing to do, and I'm losing time and money. Who's winning in this scenario? Nobody! Yet all of this could have been avoided with a meeting and some clear communication before the house was framed.

The examples I've just given you are fictional, but they are not far from the truth. If you've been in the field for awhile, you know what I'm talking about. Lots of problems crop up during jobs. Remodeling work is often plagued with unexpected trouble. Even new construction can throw you curves. If you find yourself wrapped up in conflicts like those in these examples, you need not feel helpless. A little understanding of what your plumbing contractor has to deal with can give you an edge. I'm not going to attempt to make you into a plumber, but I do plan to get you thinking like one. If I'm successful in doing this, you will be a much happier contractor.

Honing communications skills

Strong communication skills are essential to total success in business. If you can't communicate with your subcontractors, it is unreasonable to believe that you have any chance for real success. Talking with your subs before work is started can eliminate problems before they ever exist. Most contractors know this, but few do anything about it. I'm constantly surprised at the number of general contractors who don't want to be bothered with talking to me about upcoming work. It is common for contractors to wait until the last minute before consulting the plumber. This makes my job more difficult and typically increases the cost for my services.

I run into many little problems when plumbing new houses, but a majority of the serious problems pop up on remodeling jobs. I can't tell you how often remodeling contractors have called me in as a plumber, yet never afford me the opportunity to see the job before I begin. Contractors typically plug in a price for plumbing when giving an estimate and wait until they have acquired the job and are ready for the plumbing phase before calling in a plumber. This is just plain stupid. There are so many potential problems associated with plumbing and remodeling work that any contractor who gambles on this phase of work could be creating a personal nightmare that could be extremely costly. Allow me to illustrate.

Let's say you are a remodeling contractor. A customer wishes to hire you to convert an attic into living space. The new space will contain two bedrooms and a full bathroom. You work up an estimate using standard cost guides and quote the customer a price. A few days later, you are awarded the job. For whatever reason, you never called a plumbing contractor in to look the job over before the paperwork was signed. You just trusted the figures given in your cost guide. Now you have the job under contract, and you're committed to your price and work description. Are you ready for a big surprise?

For the purpose of this example, we will say that the existing living space in the home contains one full bathroom and one half-bath. In other words, there are two toilets in the existing home. Your work requires you to add an additional bathroom in the converted attic. This means adding a third toilet. Did you know that major plumbing codes will not allow more than two toilets to discharge into a 3-inch pipe? Did you happen to check the size of the building drain before you gave your customer a price? What size is the building sewer? If either the building drain or building sewer in the home has a 3-inch diameter, you have an expensive problem on your hands. You may have to either run a separate building drain and sewer for the new bathroom or increase the size of the existing drains. Either way, you could be looking at a lot of excavation and site work that you never planned for in your price. Then there is, of course, the cost of paying a plumber to install the new piping.

All in all, your failure to confirm the sizes of these drains could cost you several thousand dollars. In fact, you may wind up losing so much money that you are paying the homeowner for the privilege of doing the job. Can you afford such a mistake? I can't, and I doubt if you can either.

Could something like this really happen? Yes, it can. In fact, I've been involved in jobs where this type of problem actually happened. If you catch the problem during the bidding phase, you can bring it to the attention of the property owner and allow for the additional costs in your quote. Finding out about this after the fact, however, can be disastrous. Here's another example.

In this example, you are giving a customer a price to install a bathroom in a basement. You're no fool; you know that the bathroom in this case will require a sump and a sewage pump. Going to your trusty cost guide, you work up an estimate and give it to the customer. A contract is signed and you begin work. Are you safe? That depends. Did you factor in the cost for installing a 2-inch vent up through the house and out the roof? If you didn't, you've got a problem.

When a sump is installed for a basement bath, it must be equipped with a 2-inch vent that rises to open air. In some cases the vent may tie into an existing vent at some point in the house. However, the tie-in will have to be done at least 42 inches above the highest fixture, making the connection well into the finished living space of the home. Who do you suppose is going to have to pay for repairs to interior walls that a plumber destroys during the installation of the vent? If you guessed you, you're right. See how fast everything can turn to trouble if you don't talk with your plumber before giving firm prices and starting work?

I would like to say that I'm not trying to scare you with these examples, but I am. The truth is, I want to get your attention. All of the examples I've given you could become reality on your next job. I know this to be a fact because there have been countless occasions when such problems have occurred with my colleagues. For the security of your business, be thorough when bidding your next job. You may have avoided major financial losses in the past, but the next job could cost you a bundle.

The best way to protect yourself when bidding work is to involve other people (e.g., licensed plumbing contractors). If you get prices and written quotes from three plumbing companies, it is unlikely that you will get burned. This is not to say that trouble won't still come up, but more than likely the burden of correcting the problem will rest on your plumber's shoulders, not yours.

When you bring in a professional plumber to evaluate a job, the plumber should look for code violations and installation problems that may pop up. If the plumber fails to thoroughly inspect the job and problems arise, the financial responsibility for the extra work will be the plumber's. In this scenario, you've got the leverage. This is a much better position to be in than the ones we have discussed.

Talk to a plumber before taking on a job that involves that trade. Cost guides and personal experience are wonderful, but there is no substitute for consulting with experts. Short of dozens of disclaimers in your contract with a customer, the only way to avoid financial losses is to consult your plumber before making a firm commitment in price.

Plumbers are expensive, but talk is usually cheap. Almost any plumbing contractor will meet with you on a job to work up an estimate. It doesn't cost you anything to prevail upon the expert guidance of a licensed plumber during the bidding phase, so why risk thousands of dollars by not calling someone in? Once you ink a deal with a customer, you have no place to run. If there are going to be problems with a job, you are much better off to identify them early.

Communication is your key to safety and success. I stress this point because I know communication is lacking in the business procedures of many contractors. Please take the time to talk to your plumber before you put yourself in a position to take a huge financial hit.

Planning for success

If communication is the key to success, planning is what turns the knob and opens the door. So much can be accomplished with proper planning that it is senseless to work without it. A large percentage of contractors, however, never plan their plumbing layouts in advance. Often they wait until the last minute; then modifications have to be made to accommodate plumbing. This is an expensive habit. Much of your talking with a plumber should be on the subject of planning.

I want you to think back to the examples I gave you earlier. Remember when I was forced to cut out a floor joist? This problem could have been easily avoided. If you had instructed your carpenters to lay out the joists to avoid any toilet locations, I would not have had to deal with the obstruction. It wouldn't have been there. As a seasoned contractor, you should know to do this without any prompting. As hectic as projects get, though, such communication and planning is often forgotten. Insist on meeting with the plumber before a job starts. In fact, the plumber should insist on a meeting beforehand.

For an example of how such planning and communication should work, think back to the problem of the toilet being positioned over the foyer. If you, your plumber, and the customer had discussed this problem while it was still in the blueprint stage, a decision could have been reached before it became an on-site disruption. Again, planning in advance could have eliminated the problem.

I've worked with some general contractors who have been very good at preplanning plumbing routes. Some of them have arranged to have 6-inch partitions built to accommodate my plumbing, and many of them have worked with me to avoid cutting joists. I can recall a number of times when vent locations were chosen in advance to make sure that they would not detract from the front elevation of a home. When I've worked with these contractors, jobs have run smoothly and everyone involved was happy. Unfortunately, contractors like these are a minority. The majority of contractors don't take the time to look ahead. The result is angry subs, frustrated customers, and a general contractor with a lot of problems.

If you spend a good portion of your day trying to solve problems that crop up after the fact, how much money are you making? Not nearly as much as you could. When your time is put to better use, however, you can increase your income. Problems that plague your business and require your time cost you money both directly and indirectly. The direct costs come with dealing with the problems. Indirect costs include that of losing money because you are not available to pursue more profitable ventures. Such losses can be avoided by careful planning. Planning requires a little of your time, but it can save you money and eliminate many headaches.

As I said earlier, plumbing is a logical trade. There are, however, many subtle aspects of the trade that can produce serious problems for a general contractor. Plumbing codes are filled with requirements by which licensed plumbers must abide. It's possible to obtain a variance from the code under special hardship cases, but most of the time the code is enforced.

One simple sentence in a code book can ruin your whole day. For example, suppose you were doing a kitchen remodeling job and realized, after the fact, that adding the new dishwasher required you to upgrade the kitchen sink drain from a 1½-inch diameter to a 2-inch diameter. You might just see some of your profits going down the drain. In such a scenario, you wouldn't be able to avoid the work required by advanced planning, but you can allow for the additional cost. This, of course, prevents you from losing money.

Planning the plumbing for a job can take on many aspects of the trade. For instance, there may be water lines to install, drains and vents to add, new fixtures to include, or improvements to make to existing plumbing or surrounding areas. Upgrading plumbing is often required in order for a project to meet certain code requirements. Some old plumbing can stay if it is grandfathered, but typically it should be eliminated even if there are no code requirements for doing so. Space requirements around fixtures can also cause problems if no prior preparation is taken. There are an awful lot of considerations to think over when planning the plumbing for a job. Let's take a moment to look at some of the aforementioned categories where problems might occur.

Installing water lines

Water lines don't normally present a lot of problems when it comes to installing them. Because of their small diameter, they can typically be installed without a lot of fuss. However, if freezing conditions are a possibility, some advance thought might be needed. There is also the question of which material to use for the water distribution system. Copper tubing is common and well-accepted, but it is expensive. Polybutylene is a cost-effective alternative that has many good qualities. Which material will you spec out for the water lines? This is something that you should discuss with your customer.

The method with which the water pipes are installed could have a bearing on what you pay for your plumbing work. A lazy or greedy plumber might use an ineffective layout that will increase your costs. If a plumber is making money off of you with a markup on materials, the individual might gouge you by running more pipe than is needed. Do you have any idea of how many fixtures can be served by a half-inch water pipe? If you have no knowledge of the plumbing code, you are helpless to say what is and isn't excessive in terms of pipe runs. Going over such issues in a planning session could save you money in the long run.

Adding drains and vents

Drains and vents are considerably larger than water lines. Because of their larger diameter, they can be more difficult to install. The code-required pitch per foot with drain and vent installation can also cause great problems. There are ways to reduce your cost when drains and vents must be added. Like water lines, there are cost-effective ways to eliminate material needs when designing a waste-and-vent system. Is wet venting allowed in your area? Do you know what wet venting is? Could you trim some cost from your plumbing expense by knowing this type of information? Certainly!

Just having enough knowledge to ask questions intelligently can save you money. Few plumbers are going to try taking advantage of a contractor who asks about wet vents as opposed to dry vents and circuit vents as opposed to individual vents. When you begin to ask questions of this nature, plumbers should take you seriously. Very few general contractors know a vent stack from a stack vent. If you do, you're likely to gain some respect from your plumber, and that could translate into a lower plumbing cost.

Including new fixtures

There are many types of plumbing fixtures that can be replaced or added to a home. If you were to walk into the home of a customer who lives in a rural area where septic tanks are used and were asked to install a garbage disposal, what would you say? You might say that it is no problem. Most of the time it wouldn't be, but sometimes local plumbing codes prohibit the installation of garbage disposals in homes served by septic systems. If this were the case in that customer's area, and you agreed to make the installation, you're going to have some explaining to do. Your lack of knowledge may cause you to lose credibility in the eyes of your customer. This is certainly something that no contractor can afford to do. Staying informed is your best defense against problems of this nature.

Improving or replacing existing plumbing

For remodeling contractors, existing plumbing can cause all sorts of problems. If you don't have much experience in dealing with old plumbing, you need to gather as much information about the potential pitfalls associated with it as you can. Because you're the general contractor, you're the first person irate customers are going to call when they have a complaint. To illustrate this, let me give you an example from my past.

When I first started plumbing remodeling jobs, I didn't have the luxury of experience with old plumbing. Oh, I'd fixed a lot of it, but I didn't know what to expect from a remodeler's perspective. I learned what to look for quickly, and at some cost. One of the best examples I can give you has to do with old galvanized steel pipe that was used for years in kitchen sink drains.

I was involved in several kitchen remodeling jobs in an old apartment building. This particular series of work had me installing dishwashers and garbage disposals in kitchens where such appliances had never been installed. The drains for the kitchen sinks in these dwellings were all made of galvanized steel. Trying to save a few bucks, I was converting the galvanized pipe to plastic at the point where the drain turned out of the kitchen wall. This left an elbow fitting and a run of galvanized pipe in the wall. Although the walls were opened up to allow free access to all plumbing at the time of the rough-in, I chose not to take my plastic conversion all the way to the cast-iron stack. At the time I wanted to save some money.

When I completed the plumbing in each unit, I ran water in the sink and tested everything. I even filled the sink and released all of the water at once to make sure there were no pressure leaks or problems. All my work passed inspection from the code officers, and everything seem fine, until the units became occupied.

When people moved into the building and started using their kitchens, I started getting telephone calls from the general contractor I had worked with on the job. Many of the residents were having trouble with slow-draining kitchen sinks. In some units, the sinks weren't draining at all. This seemed strange because I had tested the sinks thoroughly before leaving the job. Nevertheless, I had to respond to the complaints without charging for my time because the work was still under warranty.

When I went out on the first service calls, I snaked out the drains and left with everything working fine. A week or two later, though, the same residents were complaining again. The contractor was getting a little miffed with me. I went back out to the condos and snaked the drains again. This went on for some time, until I decided there had to be something wrong. It was improbable that several owners of the condos would be abusing their drains to a point where weekly stoppages were common. I had to make a decision on what to do. After talking with the contractor, I agreed to open up the wall in one of the units and investigate the problem.

After opening up the nice, new walls in one of the kitchens, I cut out all of the galvanized pipe. The inside of the pipe was nearly closed with rust and crud. Because galvanized pipe tends to rust, the rough spots that had formed on the interior of the pipe began to collect debris that was passing down the drain. My snake had been cutting a hole through the build-up of grease and other debris, but the opening was not large enough to accommodate the waste from the newly installed garbage disposal. Furthermore, the volume of water from the new dishwasher was too much for the restricted pipe to handle.

In the end, I wound up replacing the old piping with new plastic pipe. This solved the problem. Luckily, the contractor was gracious enough to absorb the cost of repairing the walls in the kitchens, because my contract dictated clear terms of responsibility on that job. My work had been done in compliance with the local building code and I had never agreed to replace all piping in the kitchen wall. The agreement I had made with the contractor called for all of my work to be located under the sink, thus excluding piping in the wall. This was good for me, but bad for the contractor.

After identifying the problem, I went back to all the units experiencing problems and replaced their drains. Although I was under no obligation to do so, I chose to do this at no cost to the contractor. The contractor and I both learned a lesson from this job. In all the other units we remodeled, we ripped out the galvanized pipe as a standard procedure.

What is the moral to this story? Just because there is not a code requirement forcing you to replace something, sometimes it is best to do it anyway. Going to a little extra expense in the beginning can save you a lot of money in the end. Existing plumbing can create a lot of trouble for you, so it's best to be well informed in this area. Fortunately for you, this book is going to help you with this goal.

Insisting upon quality

The quality of workmanship and materials that go into a job can set the pace for your reputation and business success. If you cut too many corners in these areas, the price you wind up paying could be severe. There is no need to sacrifice quality in your jobs. While you might not have a constant supply of jobs that require and pay for top-quality work, you should never drop below your minimum standards of doing a good job. If you have a customer who insists on having you do work that is below your level of satisfaction, don't accept the job. You are better off not doing the work than being tagged a shoddy contractor.

There are plenty of ways to save money on your plumbing installations without giving up the quality image you want to present. Many of these methods are discussed in later chapters. In fact, each of the following chapters will contribute to your ability to offer customers good work at a fair price. If you want to stay competitive and stay in business, reasonable, quality workmanship is essential.

2

Determining logical progression of events

One way to make more money on a job is to schedule your work in a logical progression. This is a simple process, but many contractors struggle with this concept and typically wind up never learning the cost-effective scheduling process at all. These contractors don't last long in the business. If you want to prosper in the building and remodeling business, you have to learn the fine art of scheduling. This is not an option, but a requirement.

How you schedule your work can have a lot to do with the number of problems you'll have to confront on a daily basis. Scheduling can also influence the amount of money you put in the bank each week. As a general contractor, a big part of your job is coordinating subcontractors and suppliers. If you are not adept at doing this, you can't reach your maximum potential.

Plumbing can become expensive when scheduling problems occur. For example, if you're paying a plumbing crew to stand around while your carpenters finish the last-minute framing for the bathtub that must go in, then you're probably losing money. Many plumbers have clauses in their contracts that allow them to charge extra if they are called to a job only to find that it's not ready for them yet. The aforementioned scenario could cost the general contractor well over $100 an hour for such a scheduling mishap. I doubt if you can afford many of those mistakes.

The cost of paying extra for work that is poorly planned is not the only consideration. As a general contractor, you have to answer to the customer. If your customer comes home during the day or visits a job site and sees crews standing around, you might get a telephone call, especially if it happens more than once. When a job is not progressing as scheduled, you may have difficulty dealing with the customer. But remember, customer satisfaction should always come first.

Calling plumbing too soon

If you call in your plumbing contractor too soon, he or she may become upset if a productive work schedule is not met. The plumber might be able to keep occupied accomplishing some needed odds and ends, but the delays will inevitably cause him or her to lose money because most plumbing jobs are bid on a full-scale production basis, not on an hourly basis. If the work has to be done in little bits and pieces, time that the plumber could use on other projects will be lost. Time is money, so you can bet this won't continue for long before the plumber is no longer working for you.

Not only are plumbers expensive, but it is hard to find a good, dependable one. Once you've found a plumbing contractor who lives up to your expectations, don't alienate him or her. Calling the plumber in before a job is completely ready is a good way to do just that. It's like the little boy who cried wolf. If you call your plumber too often for jobs that aren't ready, someday your call will go unanswered.

Many general contractors don't have a good grasp of what plumbers need before they can do their jobs efficiently. Clear communications between you and your plumber is the easiest way to determine what is needed. Many contractors feel that their plumbers should jump right into a job, regardless of whether it's complete. Established plumbers typically don't work under such adverse conditions because they are not as profitable. Chaos reduces your productivity, which in turn reduces profitability. You must keep this in mind when scheduling your plumber.

Bringing plumbing in too late

If you wait too late to bring in your plumbing contractor, you could be setting yourself up for a host of problems. Some of the potential trouble can get very expensive. Can you imagine pouring a new concrete floor and then having to tear part of it out so a plumber can install piping? I can remember a townhouse development where this happened. Most situations are not as extreme, but there have been worse cases.

Even if you don't have to rip out a new slab, calling the plumbing contractor later than you should can still get expensive. Sometimes the cost is in destroying work already done by someone else. The cost of labor charged by the plumbing company might be the only visible sign of damage, but at the rates plumbers charge, the damage can be crippling to your job cost.

There is more than money at stake when you don't schedule your subcontractors properly. Oftentimes individual subcontractor scheduling problems can lead to delays with several other contractors. To ensure a good reputation for your company and a satisfied customer, you have to meet your commitments. Good management of your subcontractors will also help keep your bank account growing with each job. Because it is so important to schedule your plumbers properly, let's take a look at what signals you to bring in a plumber.

Scheduling plumbing in new construction

When you are involved in new construction, your timing for bringing in a plumbing contractor is different from what it might be on a remodeling job. Typically, it is much easier to schedule plumbers, and other subs, for new construction. When you are building something from the ground up, there are generally fewer surprises than when you are doing renovation or remodeling work. Unless you experience unusual circumstances, the production schedule I'm about to give you for plumbers should work well. In my examples I use residential construction as my basis. If you're concerned about commercial work being different, don't worry. I make reference to differences between residential and commercial work as needed.

Obtaining bids

The first step in working with plumbing contractors is obtaining bids for upcoming jobs. You should involve plumbing contractors early in the bidding stage. If you are bidding from blueprints and specifications, this is a simple process. You simply send the documents out to various plumbing companies and wait for their proposals. The plumbers may need some background information pertaining to the site, such as the location of water and sewer connections, but typically a physical visit will not normally be required with new work. If it is, the general contractor should coordinate the visit.

Before we go on, I want to clarify what I'm calling new work. By new work, I'm referring to the construction of a new building on a site where there has not been a building. The construction of a new room addition, under my descriptions, would be classified as remodeling. Even though the work involved with building an addition is very similar to new construction, it differs from building a complete building from scratch.

Holding an office meeting

I believe that an office meeting should be held with the plumbing contractor you plan to hire before any work is done in the field. This meeting will give you a chance to review all aspects of the plumber's bid and the work description. Ideally this should be done before a price is presented to a customer. At the latest, you should hold this meeting before any foundation work is completed. A lot can come out of a face-to-face meeting with your plumber.

When you meet with your plumbing contractor, you can lay out a viable work schedule for all phases of the plumbing. This is the time to find out when your plumber would like to be sent in for different aspects of the job. For example, let's say that your plumber will be installing all plumbing to a point that is five feet outside of the building's foundation. This is a common approach to how plumbers bid jobs. Does the plumber want to be notified prior to the installation of footings or doesn't this matter?

Some plumbers go in after a foundation is made and tunnel under it to get their building sewer out. Personally, I prefer to go in before the footing is poured. This allows me to dig more easily. I can then install a sleeve that will be covered with concrete when the footing is poured. By doing this, I eliminate the back-breaking, time-consuming work of tunneling under a new footing. The job can be done either way, but the cost is going to be higher if you have to tunnel under the footing. By using a little advanced planning and installing the sleeve before the footing is poured, you can reduce your plumbing costs.

If a building is going to have a basement, the water service and sewer will probably be installed so that it penetrates the foundation wall. That being the case, if a sleeve has not already been put in the footing, you can use a jackhammer to make access for the pipes at that location. The best practice is to install a sleeve prior to the concrete. Even if the foundation is being built by masons, providing a sleeve in the wall is the easiest and most inexpensive way out. It only makes sense to save time and money everywhere you can when it doesn't affect the quality of your job. Installing sleeves for all wall and footing penetrations also makes for a neater job as well as a lower job cost.

When you complete your office meeting, you should have a tentative schedule hammered out with your plumber. While you may not have actual dates penciled in, you will know at what stages the plumber wants to be notified. A meeting like this doesn't take long and the rewards can be almost immeasurable.

Making your connections

Depending upon the availability of utilities, there will be water and sewer connections to be made when a new building is constructed. If you're working in a rural area, these connections may be replaced by a well and septic system. Either way, you need to be in tune with your plumber on these issues. Connection locations and the routes of pipes should be determined before major site work is done and the foundation is installed. Because these connections may influence the location of driveways, parking areas, and so forth, the mechanics of this work should be reviewed before physical field work is started.

Installing underground plumbing

Underground plumbing is a term that could apply to water and sewer lines, but it is usually confined to mean plumbing that is installed under concrete floors. Many plumbers refer to this phase of plumbing as the groundworks. Most plumbers are willing to do the groundworks at any time after the foundation is built and before the floor is poured. Very few do it before a foundation is installed because it is just too difficult to pinpoint wall locations if there is not a foundation to use as a benchmark.

Exactly when should your plumber be called in to do the underground plumbing? This depends a little upon individual circumstances. Typically the plumbing contractor will come in after the foundation walls are up and before any grading for the slab has been done. This means that there will not be any reinforcing wire and gravel when the plumber installs the groundworks. There may, however, be occasions when a plumber will want the gravel installed prior to the installation of underground plumbing. This is something you will have to sort out with your plumber. Basically, plan on scheduling the groundworks to be done after the foundation is in and before the existing dirt is disturbed or covered on the inside of the foundation. A smart plumber will want to have sleeves installed before footings and foundations are created, so keep this in mind when scheduling your crews.

Completing rough-in work

Rough-in work involves the installation of water pipes, drains, and vents that will be concealed as a building is completed. This is a phase of plumbing where many contractors make enemies of their plumbers. It is usually a matter of planning, or the lack thereof, and scheduling that sours the relationship between the general contractor and the plumber. There is no rule that says when a plumber should do rough-in work, but it must be done before walls and ceilings are closed in with drywall. This being the case, what subcontractors should be called in first for rough-in work?

There are three primary rough-in subcontractors: heating and air-conditioning, plumbing, and electrical work. So who goes first? Which sub should come in second? This is where being a general contractor can get a little complicated.

General contractors often make mistakes in choosing which subs to call in first. Normally, this is not a matter of inexperience or lack of knowledge. Most generals know which subs should have the first opportunity at a rough-in, but circumstances can slant their decisions. For example, a heating contractor may not be available when a job is ready for rough-in. If this is the case, a contractor may authorize a plumber to go in to complete his or her installation in front of the heating contractor. This is done to avoid lost time and to expedite payment schedules. Allowing the plumber to work first can create some significant problems though.

If a house is going to have ductwork installed in it, either for heating or air conditioning, this work should be roughed-in before the plumbing and electrical work. Ducts are difficult to deal with, due to their size, so the installers should have the first shot at finding suitable space for running their materials.

After ductwork, drains and vents are the most difficult elements to rough-in. With this in mind, plumbers should be allowed to install their pipes after the ductwork. The electrical rough-in work should come last because the wires are much easier to run.

Scheduling the rough-in inspection

The plumbing rough-in inspection is a part of the job that you, the general contractor, should not have to coordinate. Setting up the inspection with the local code enforcement officer is the plumbing contractor's responsibility. This doesn't mean, however, that you can shut your eyes to the need for an approved plumbing inspection. You don't have to schedule the inspection, but you have to make sure the rough-in has been approved by the local plumbing inspector before you take any steps to conceal the work.

There are many types of inspections required within thc plumbing trade, including: water services and sewers; groundworks; and water pipes, drains, and vents. All must be tested, inspected, and approved before they can be buried. If you, as a general contractor, cover up plumbing work that has not yet been inspected and approved, you may be forced to remove whatever was installed that is concealing the plumbing. Make sure you have approved inspection slips on file before you conceal any plumbing work.

Setting fixtures

The final phase of plumbing involves the setting of fixtures. This particular part of a job can get confusing. You can find yourself wondering whether the flooring contractor should go in before the plumber or if the plumber should go in before the flooring contractor. Does an electrician wire a dishwasher before a plumber plumbs it? At what point should a plumber be called in to set fixtures?

Plumbers like to set all of the plumbing fixtures for a job at one time. This means that counters and countertops must be installed before the plumber arrives. It is usually considered best to have finished flooring installed in bathrooms before plumbers set their fixtures. It is possible to install vinyl flooring, and other types of flooring, after a plumber has set vanities and toilets, but it is neater if the flooring goes in first. Garbage disposals, water heaters, and dishwashers should be installed before electricians are called in to wire them up.

After all plumbing fixtures are installed, a code inspection, performed by the local plumbing inspector, is required. Once the installation passes, you are done with your plumbing duties.

Taking on remodeling jobs

Remodeling jobs that involve plumbing can present scheduling difficulties for general contractors. Schedules that worked in new construction may not pay off in remodeling because there are more unknown elements to deal with when working with existing conditions. This can require you to have plumbers on the job at odd times and without a perfect schedule from which to work. Let's examine why "on-call plumbers" might be needed with remodeling jobs.

Conducting site visits

Site visits are very important when working with remodeling and renovation jobs because existing conditions can create a variety of problems for plumbers. While new construction can be bid from plans and specs with good accuracy, remodeling work requires a physical inspection if a solid bid is expected.

Checking water and sewer services

Water and sewer connections for remodeling jobs are usually in place and in service; therefore, there is not normally any need to install separate service lines. There could, however, be such a need if the existing pipes are too small to handle the increased load of the space being remodeled.

When you conduct a site visit with your plumber, you can ask if the water and sewer lines might possibly need to be modified. Nine times out of ten, no modifications will be necessary. At this point, you are ready to move on. If for some reason the existing water and sewer services are not adequate, you may have a lot of unexpected work in front of you. This is why it is a good idea to insist that the plumbers who are bidding your jobs make all bids based on installing and altering work while complying with the local plumbing code.

Scheduling efficient rip-outs

Rip-outs, the disconnecting and removing of old fixtures, are a frequent part of a remodeling job. Depending upon the circumstances surrounding a particular job, rip-outs can create friction between contractors and homeowners. For example, if you have your plumber remove a kitchen sink too quickly during the remodeling process, the customer might be left with no sink for some time. It would be even more incredible if the crew removed the toilet too soon in a job. Scheduling efficient rip-outs is one of the many strong points an experienced remodeling contractor has to offer.

When and how you schedule rip-outs depends on your specific type of work. For example, in the case of kitchen remodeling, you should leave the sink in operation for as long as possible. Similarly, if a home only has one bathroom and you are gutting it, you must make sure that your plumbing schedule is tight and on target. To accomplish these goals, you must choose a dependable plumbing contractor and maintain clear communication with that plumber.

Once you have picked a good plumber, keep the work flowing. This will keep the plumber happy and put you in a good position to ask the plumber for a favor when needed. For example, you might need a plumber to come into a job just to remove a toilet, a 15-minute job for an experienced plumber, so that new flooring can be installed. This type of special service is not uncommon. As a general contractor, though, you should have a lot of work to offer that plumber so that he will work with you in special circumstances to get the job done. Using work as a lever, when needed, may seem a little harsh, but it allows you to meet your deadlines. Most plumbers are willing to work closely with the general contractors for whom they work.

Scheduling efficient rip-outs requires that the person in charge be aware of the optimum time to make the removal happen. For example, assume that you are doing a kitchen job and the dishwasher and disposal need to be removed. Should you have an electrician come in first or a plumber? You can probably get by with whichever subcontractor is available first. However, getting an electrician in first makes sense. Having the wires disconnected when the plumber arrives will make the job go smoother. Many plumbers are willing to make electrical disconnections themselves. If your plumber is willing to do this, you won't have to be bothered with bringing in an electrician for the rip-out.

Completing unexpected work

Unexpected work often comes up in remodeling jobs. When plumbing is involved, you need a plumber who will come when you call. Because unexpected work can't be planned, you have to roll with the punches. This is easier to do if you have an ongoing relationship with your plumbing contractor.

What types of work might be required of your plumber between scheduled work phases? Examples include removing an existing pipe to allow for the installation of a new window or door, replacing a fixture accidentally damaged during construction, or providing quick service because a nail hit an existing water pipe during construction. Unexpected plumbing duties are not part of every job, but they happen often enough that you should have arrangements worked out with your plumber to handle such emergencies and unscheduled service needs.

Scheduling rough-in work

The rough-in work for remodeling jobs should follow the same schedule as for new construction. Ductwork alterations and installations should be done before a plumber is set into motion on rough-in work, and the electrician should take over after the plumber is through.

When doing remodeling work, crisp schedules don't usually work as well as they do with new construction. There may be many times when trades are overlapping each other. For example, an electrician might be doing rip-out work while a plumber is doing rough-in work. As long as the trades are not getting in each other's way, this is fine. However, when it comes to installation work, the primary schedule should be followed.

Reaching the final plumbing stage

Once you reach the final stage of plumbing in a remodeling job, the scheduling process can be compared to that of new construction. There shouldn't be any problems associated with plumbing at this stage in a remodeling job that wouldn't be encountered with new construction.

Keeping to your production schedule

The use of a production schedule is very helpful in construction. Scheduled dates on your production plan may come and go without being met, but just having an organized plan will help you juggle your subcontractors more effectively. If you put a production plan down on paper, you may discover crucial steps that have been previously overlooked. For example, when should you have bathtubs and showers delivered to your job site? If you wait too long to schedule their delivery, you could find that they cannot be put in the building without destroying completed carpentry work.

One-piece tub-shower combinations are so big and bulky that they often won't fit through small doors or up interior stairs that have been drywalled. In order to not run into problems, you have to get these units on the job early in the framing stage. Failure to do so will mean tearing out wall studs and such. As simple as this little matter is to handle, many contractors neglect to order their tubs with ideal delivery dates. Your plumber might be responsible for the timing and actual delivery of the tubs, but it is normally the responsibility of a general contractor to approve the delivery date.

With new construction, getting a shower-tub combination in at the right time is just a scheduling issue, but trying to get a one-piece tub into a remodeling job is pretty near impossible. There are exceptions, of course, but the mechanics of getting a one-piece tub-shower combination up existing stairs or through interior doors does not leave much hope for success. If you will be doing work where a large opening is available, such as in the case of some attic conversions, one-piece shower-tub units can be used. They can be passed through the opening relatively easily. It's the scheduling of the delivery that is most important.

Installing standard bathtubs

The use of standard bathtubs and sectional tub-shower units relieves you of most complications experienced with one-piece units. The only real trick is that the bathing unit must be set in place before drywall is hung. Failure to do this will result in cutting and patching newly-hung drywall to get coverage over the nailing flange of the unit. This mistake is rarely made, but I'm sure it has probably happened to someone.

Adding whirlpool tubs

Whirlpool tubs, spas, garden tubs, and soaking tubs can all present a contractor with size problems. Some of these fixtures are too large to fit through interior doors. This means that the units should be set in place during the framing stages of new construction. In the case of remodeling, it is essential that a contractor make sure access is available for oversized tubs before a commitment to install one is made.

I've been involved on more than one job where the tub specified by the customer would not fit into the space allotted. There have also been times when a specifically chosen tub could not physically get to the desired location without serious destruction of existing conditions, such as interior doors. If you or your plumber don't catch these problems before you are committed to doing the work, the out-of-pocket costs can rob you of any profit you had hoped to make.

Upgrading the water heaters

Water heaters may not seem like much of a threat to you or your job, but they can be. If you've been hired for a remodeling job where additional plumbing fixtures are being added, it may be necessary to upgrade the existing water heater. A larger heater may be required to handle the additional load of new fixtures. First of all, you have to account for the cost of the heater and its installation under these conditions. Secondly, you have to make sure there is some place to put the larger heater. It is not uncommon to find condos and houses where the water heater being used is the largest size that the storage space can accommodate. Needing a bigger water heater and not having a place to put it is a real problem.

Preplanning plumbing to avoid problems

Preplanning all aspects of plumbing for your jobs is the only way to avoid problems and cost overruns. Even with solid planning, there are going to be times when something will happen that you weren't expecting. You will have more than enough trouble with plumbing under the best of circumstances, so leave as little to chance as possible.

It is easy to get caught up in the bidding process and fail to see danger signs. You can't afford not to think about oversized tubs, fixture-spacing requirements, available space, access, and other considerations. If you rely solely on your plumbers, you might not have to bear all of this burden, but you will still be right in the middle of trouble when it happens. As the general contractor, you can't step outside the circle and point your finger. Your customers are going to hold you accountable. The fact that your plumber has to pay for a mistake helps a lot, but it doesn't make your customer any happier. You have to work closely with your plumber to avoid all the problems that you can. In order to do this best, you have to possess at least a general knowledge of the plumbing trade and plumbing procedures. Pay attention to what you learn from this book and apply it in the field and in your office to minimize your risk of costly and embarrassing mistakes.

3

Coordinating subcontractors and employees

The cornerstone of success for general contractors is being able to efficiently coordinate subcontractors and employees. Until you learn how to do this, you can never reach your maximum profit potential. Is this coordination process simple? No. If being a general contractor was easy, it wouldn't be such a lucrative career path, and not so many people would fail. In reality, taking responsibility for a bunch of independent contractors is far from simple or easy. If you've been a general contractor for awhile, you know this is true.

When we talk about subcontractors, we're referring to independent contractors. This group of people live up to their name at times. They can be very independent and difficult to control, and if you lose control of your subcontractors, you've lost control of your business. You can't afford to do that.

Before you can efficiently coordinate subcontractors, you must be in control. If you're not, you're just wasting your time. A general contractor often employs many methods in an attempt to gain this control over the subcontractors. Most of the methods used relate to money, although some of them are not as direct as withholding a payment or two.

Setting the ground rules

One of the first things you should do when establishing a working relationship with a subcontractor is set ground rules. Although plumbing contractors are discussed in this book, most of the principles introduced can be applied to other independent contractors. A plumber is one of the most independent subcontractors you'll hire. Because there is a great demand for good plumbers, people with marketable plumbing skills don't have to look very hard to find work. This makes it tough on general contractors to keep plumbers at their beck and call. But there are ways to gain control and respect at the same time. If you learn these methods and use them properly, your plumbers will go out of their way to take care of all your plumbing needs.

Part of your efficiency as a general contractor depends on having subcontractors you can count on in times of need. If you find yourself in need of a plumber on short notice, can you be sure that your regular plumbing contractor will take care of you? If you're not confident of this, you and your plumber need to have a talk. You may even need a new plumber.

I learned a long time ago that the best time to work out the details of what is expected of subcontractors is right up front. If you have a meeting with your subs where the topic of conversation revolves around your future expectations of them and their job, there is a good chance that you'll be able to pick out the more cooperative ones early on. There is no guarantee of this, but the odds are in your favor. Certainly, a meeting of this type is worth your time and effort. Anything you can do to cement a solid working relationship will benefit you as future jobs come along.

Many subcontractors are suspicious of general contractors in the early stages of a working relationship. As a plumbing contractor, I can tell you that some general contractors don't pay their bills on time or at all. This is a great concern to most subcontractors. Because generals rarely authorize advance deposits on work to be done, subs are at risk. Unless you have references from several of your existing subs for new subcontractors, you may not have as much power as you think.

A lot of general contractors who are new to the business feel that they hold all the cards. They think that without general contractors, subcontractors have no way to ply their trade. This simply isn't true. A good plumbing contractor doesn't need a general contractor to create a successful business enterprise. Having quality general contractors to work with is certainly a plus, but every contractor knows that his or her survival doesn't depend on work from a general. This fact is one that some generals don't want to or can't accept. The relationship between a general contractor and a subcontractor has to work both ways to be successful. This is never more apparent than when a general needs a special favor from a sub, and believe me, there will be times when you will have to ask for special consideration.

As a general contractor, you should be the person in control of your jobs. Most subcontractors accept this notion. However, the methods that you implement to gain and maintain control can be your downfall. If you attempt to make major power plays, you will lose your best subs. Independent contractors who are hurting for work may put up with your dictatorship for awhile, but most of them will leave once they are in a position to do so. Top-notch subs won't stand for much abuse. They will simply tell you what you can do with your work and walk away. With this in mind, you have to develop tactful ways to work with your subcontractors. It is all part of building the efficiency of your company.

If you're changing subcontractors frequently, you're doing something wrong. Either you're not screening the subs well enough when you hire them or you are treating them poorly. Either way, you suffer financial losses. It takes time to round up new subs, the time you could be spending doing more lucrative things. To alleviate the lost time, put your expectations on the table with subcontractors before you bring them into your business circle.

Establishing your requirements

What types of requirements will you insist upon with your plumbing contractors? Naturally, you'll want the minimum standard, quality work at a fair price, but there are other issues that are just as important as pricing. For example, will the plumber guarantee you 24-hour service? How long will you have to wait for a response if you page your plumber? Are your emergency calls going to be taken by an answering machine, a secretary, or a human answering service? Does your company require all subcontractors to have cellular phones for immediate access? These are just some of the questions that you have to sort through when setting the guidelines for subcontractors who serve your company.

Plumbing is a trade where almost anything can and will happen. How about these examples? A compression ferrule fails and water overflows the upstairs bathroom, leaking water through the ceiling of the first floor; a soldered joint breaks loose in the middle of the night, causing major flooding; or the bathroom drains become clogged in the middle of a cocktail party. You name it, and it can happen. How will you deal with these situations when your customer calls you at midnight in a panic? If you don't already have an answer in mind, you had better figure one out. There is no way that any general contractor can be active with several jobs and not experience some unexpected plumbing problems.

Choosing competent subcontractors is one of the best ways to reduce the risk of late-night calls. Even the best of plumbers, however, can't prevent all forms of plumbing disasters. Sometimes things just happen. This is when you need to know that you have a quick-response plumber only a phone call away. Without such an arrangement, you could be in deep water with your customers (no pun intended).

Because I am a plumbing contractor, I don't typically hire plumbers. When I have contracted independent plumbers, however, I've always set forth a company policy that they were expected to follow. Because you will be dealing with independent plumbing contractors on a daily basis, and with my general contracting business I don't, I will share some of my past experiences with other general contractors to make you aware of the types of requirements they have placed on me as a plumbing contractor.

Living through the toughest

The toughest requirements I've encountered were from a general contractor who always had to be in complete control. His idea of a good working relationship was, in my opinion, unreasonable. This contractor wanted both my pager and home phone numbers. I didn't mind giving up my pager number, but I fought the idea of relinquishing my personal phone number because I knew that if the contractor needed me in an emergency, my answering service would call or page me. The contractor also insisted that I maintain a mobile phone, and this was back when mobile phones were very expensive. As it happened, I had a very expensive mobile phone (it cost me $1800), but I didn't wish to make the number available to general contractors, partly because of the high cost of air time. But more importantly, the phone was purchased and operated for my convenience, not that of others. I may have been a bit too independent.

In addition to all the instant contact that was expected, the general contractor also had some stringent plumbing guidelines to follow on the job. During my second meeting with the contractor, I was handed a field manual that outlined the instructions to be followed on all jobs. Much of what was in the manual was reasonable, but there were a lot of outrageous instructions. For example, this contractor went so far as to specify how far from a wall a chrome nipple could extend before a stop valve was installed. There were many other picky points in the manual. It didn't take me long to determine that working for this contractor would be very demanding, to say the least. I chose not to work for his company.

Choosing a contractor

Other contractors that I have met with have fallen into various categories. Some have been so slack in their direction and organization that I decided not to work for them. Trying to please someone who doesn't know what it is he or she wants is just as hard as working for someone who is going to give every job the white-glove test. I've settled in with contractors who are somewhere in the middle of the two extremes and everything has worked out about as well as can be expected.

I believe that the tough contractor I told you about first had the right idea of how to run a contracting business. His field manual made every aspect of a subcontractor's responsibilities known. There was no room for confusion. If the manual had been tempered to a more equitable form, I would have welcomed it. Having subcontractors available on short notice is fair, especially if the requirement is to cover warranty work that has gone bad. As a general contractor, you have to pick a path that you are comfortable with when setting rules for your subcontractors, and then give them some guidelines to follow.

Setting my rules

My way of dealing with subcontractors is similar in form to the methods used by the tough contractor in the first example. I require phone and pager numbers where I can get a fast response from the subcontractors. It doesn't have to be their home phone, but I want some way of contacting them quickly. My rules require all subcontractors to return any urgent message I leave within 15 minutes. For regular calls, subs have up to an hour to get in touch with me, unless I've specifically told their answering services that the return call can wait for several hours.

I don't have a field manual that I hand out to subs, but I do have a list of rules that I go over with them. Most of my rules are covered in the contracts that I use. One such rule is that all subs must clean up their work area after each day. For example, if I were hiring a plumbing contractor, I would stipulate that the plumber sweep up all wood shavings after drilling holes. Other rules that are covered in my contract include acceptable working hours. This clause is especially important with remodeling jobs where customers are living in a house while it is being remodeled. My customers might not appreciate an early-bird plumber who starts work just as the sun is coming up. Conversely, they may not wish to eat dinner while listening to the roar of a right-angle drill. My clause on work hours lets everyone, customers and subcontractors, know when work will be done.

You can set your subcontractor requirements however you like. Once you have a procedure for subcontractors to follow, you have a better chance of using them efficiently. These same principles apply to employees. Employees are easier to control, in most cases, than independent contractors. However, employees deserve to know the rules before they are hired. Once you have your work force put together, with either subs or employees, you are ready to get your field crews rolling. If you want to run your business with minimal maintenance and few breakdowns, you have to occasionally check on your workers. If you just put your people in motion and move onto other tasks without monitoring what is being accomplished, you're asking for trouble.

Scheduling your plumbers

The scheduling of your plumbers sets the tempo for your job. If you don't have a schedule, who knows what will happen? When you have a schedule that is unrealistic, everyone will become frustrated. You need to develop a viable schedule for each of your jobs. Things will go wrong from time to time, but this is to be expected. A contractor who can't work through adversity isn't going to last long in business. All it takes is one mistake from almost anyone to throw an entire schedule out of whack. There may be times when you feel that making a schedule is a waste of time because something always seems to change, but your best bet is to amend your schedule and keep on going.

If you are a general contractor who has decided to hire plumbers as employees, you might have to worry about how to keep them busy so you don't lose money. One big advantage to using subcontractors is that you don't have to pay them to sit around waiting for work to become ready. With employees, you have to maintain a fairly constant stream of work to keep the people on your payroll. Most general contractors typically are better off using subcontractors than employees for their plumbing needs.

Assuming that you have good plumbing crews available, your next biggest concern will be planning and scheduling the delivery of material to the job. If you're using subcontractors, they may handle all aspects of the plumbing for your jobs, including material deliveries. A lot of general contractors, however, prefer to buy their own bath units. If you are in this group, you must coordinate the delivery of these units with your framing crew and plumber. I discussed in the last chapter how problems arise when bath-shower units are not delivered early enough. If this responsibility rests on your shoulders, you cannot afford to overlook the delivery schedule.

The last chapter gave you a good sense of when different phases of plumbing should be done. You can refer back to that information when working up your rough schedule. Keep in mind that the suggestions I give you are general in nature and that you should consult with your own people to perfect a schedule.

Putting a schedule on paper is easy. However, seeing that work gets done in accordance with a schedule is not so simple. Planning and keeping your work on schedule is the true test of a general contractor. Coordinating all the elements of a job so that the work runs smoothly takes experience, not to mention dependable field crews and a little luck.

Holding the line

Holding the line on your overall production schedule can be extremely difficult. Although problems may not arise in the planning and executing phase of plumbing, if your other employees or subcontractors have not done what they were supposed to before plumbing arrives, the production schedule can be thrown off. This, in turn, nullifies any hopes of keeping your plumbing work on schedule. To avoid common problems that can shut your plumbers down, you or your field supervisor should make sure jobs are ready for your plumbing crew before they show up for work. This will alleviate possible problems with scheduling and keep you from losing money due to scheduling difficulties.

If you know what to look for, you can make sure jobs are ready for your plumbers in advance. Take the time to investigate the status of jobs before sending your plumbing crews in. By doing so, you will greatly reduce the risk of plumbing slowdowns and shutdowns. This is a key step in coordinating your crews. Keep in mind also that if your plumbers fall behind in their schedule, other contractors will be affected.

When plumbers are running behind, the electrician and the insulators for the job will have to be rescheduled. This, in turn, will push the rough-in inspections for plumbing, electrical, and insulation work back, affecting also the general building inspection. All of these changes force you to put off your drywall delivery and the work of your drywall hangers and tapers. Being just one day late with your plumbing can force you to make changes with all of the subcontractors, inspectors, and deliveries mentioned. And it can get worse.

Let's say that your plumbing crew misses the scheduled target date by two days. Will this make your whole job two days late? Well, unless you gain some ground in some other phase, yes. Unfortunately, a delay of two days with your plumbers could force the job to be not just two days late, but it could conceivably make it more than a week or two late. Why is this? Once you call your other subs, like your drywall contractor, and extend their start dates, you are opening up a situation where you might be told that the subcontractors can't shift their start dates without affecting other jobs. You could wind up having to wait for some time to get the subs back once you have prevented them from going into your job on schedule. This can create damaging results with your customers and your cash flow. I'm not making this scenario up, I've had it happen many times. One or two days can make a huge difference in the success of your job. Therefore, it is imperative that you set realistic schedules and stick to them.

Predicting the unpredictable

What should you look out for that may delay your plumbers? There are a multitude of possibilities for delays in the plumbing work done on a job. Some of them are unpredictable and unavoidable. Most of them, however, can be caught early enough so that they don't disrupt your schedule. The key to staying on schedule is finding and correcting the problems before your plumbers are on the job. You, or someone you trust, must make periodic inspections well before the plumbers come to a job. The following are some plumbing aspects that could cause plumbing delays.

Tapping connection sites

Connection sites for water and sewer services are one of your first concerns for the plumbers on a job. If your plumbers will be tapping into city utilities, a tap fee must be paid. Has the fee been paid? Are you paying the fee directly or is the cost included in the plumbing contractor estimate? Once you have established who is paying for tap fees, you can make sure they are paid. If the plumbing contractor is responsible for them, require him or her to send you a copy of the paid receipt. This will allow you to mark off one category of your plumbing checklist.

Some jobs are served by septic tanks rather than municipal sewer services. Is the tank going to be installed before your plumbing rough-in is done? If not, who is going to make the connection between the tank and the building sewer? If your plumber has to make a special trip back to the job to make such a connection, you will probably be charged for the extra trip. The plumber couldn't wait until the final plumbing to make this connection because, at that point, the grading work around the building would have already been done.

Most plumbing contractors bid jobs on the basis of extending the building sewer out of the foundation for a maximum length of five feet. If the lateral for the connection is installed within this distance, the plumber should be willing to hook up to it. You can't safely assume, however, that the plumber will run a longer distance to connect to a septic tank or sewer. Someone has to install the sewer from its connection point with a septic tank or sewer main to a point close to the building being served. Who is going to do this? Is the cost for this work included in the bids you have gotten? If the hookup was included in the bid price, was excavation and grading also included? Someone has to dig the trench, backfill it, and grade it. All of these are factors that can create unexpected costs for inexperienced contractors. Each can also slow down the production of a job if they are not anticipated.

Hooking up water services

The same basic questions and concerns that we have just discussed (e.g., tap fees, road cuts, excavation, backfilling) can also apply to water services. If a water service is not going to tie into a city water main, where will the water come from? It will probably come from a well. Will the well be ready for the installation of a pump during the rough-in phase of plumbing? Who will be supplying and installing the pump, the well driller or the plumbing contractor? Who's going to supply and install the pipe for the water service from the well to the building? Get these questions answered before you set a rigid plumbing schedule.

Installing sleeves

If your job will require plumbing to come under a footing or to penetrate a foundation wall, and almost any job that you do requires a new sewer or water service, you will have to decide how and when the sleeves will be installed. The easiest way is to install them before concrete is poured and masonry work is completed. This part of the job is the plumber's responsibility, but you can ensure that this part of the job is done when needed if you address this issue with the plumber early on.

Shooting the grade

Will there be enough grade for a new sewer? In other words, can you get 12 inches of ground cover over the sewer where it leaves the building, maintain a standard grade as the sewer is installed, and still have the pipe be high enough to make the connection at either a septic tank or sewer main? Most builders assume this will not be a problem, but there are times when it is, and a very big problem at that. You or your plumber should shoot a grade on the septic trench to make sure there will be enough slope to make connections at each end of the sewer while maintaining a fall in accordance with your local plumbing code.

Installing groundworks

Before you schedule plumbers to install groundworks, you must be sure the site is ready for them. This typically means ensuring that all foundation walls and piers are in place. Although some plumbers will work from string lines, most won't. It's not enough to have just the exterior foundation walls up. The piers for support columns should also be installed so that the plumber does not run pipes right through the area where the pads are needed.

Another consideration with groundworks is fill material. If your plumber is forced to install pipes in rocky ground, there should be fill material, such as crushed stone, sand, or dirt, available to support the pipes. A plumbing contractor typically will not install groundworks and leave the backfilling and support to other people. Pipes can get broken or slopes changed during the backfilling process if not monitored closely. Good plumbers will want to handle this part of the job themselves, but they will expect you to have fill material available.

Completing inside rough-in work

Inside rough-in work for plumbing can be done in various ways, but there are certain restrictions. For example, plumbing codes require minimum spacing between fixtures. If your carpenters frame an area too small or put floor joists in the way, the wood is going to have to go. Access and clearance is also an issue. If access is not available, the plumber can't do much until the matter is resolved.

Bathtubs and showers are installed during the rough-in phase. If tubs are not on-site, a plumber can't complete the rough-in or call for a code inspection. When it is your responsibility to get tubs on the job, you must do so in advance. Although getting the tub is a concern, the majority of your concern during the rough-in stage will be related to carpentry work. Are there any floor joists under toilet locations? Are all partition walls installed so that plumbing drains and vents can be run? Some plumbers refuse to rough in two-story homes where the steps have not yet been installed. This is something that you should keep in mind. It's easier for everyone on the job if steps are in place, even if the treads are temporary.

A plumber may leave the job before it is finished if framing is incomplete and inaccurate. If this happens, getting the plumber back can be an uphill battle. Check your carpenters' work closely to make sure room dimensions are the same as those shown on the working drawings. If a powder room is even one inch too narrow, your plumber can't legally install a toilet rough-in. To promote efficiency, all framing work should be done before the plumber comes to a job.

Doing final plumbing

Final plumbing is a time when many plumbers get frustrated. For example, they might get to a job to put in a kitchen sink only to find that the counter hasn't been installed yet. The sink obviously can't be installed. Or the plumber might start installing a one-piece, integral-bowl vanity top supplied by you, the general contractor, and find that it is defective, again bringing the work to a halt. If you are supplying any materials, such as tubs and vanity tops, check them for damage when they are delivered. Don't waiting for your plumber to discover the damage, which would cause even more trouble.

As a plumbing contractor, I've been sent to jobs many times when the final flooring had not yet been installed. Because most general contractors want the floors finished before plumbing fixtures are set, the plumber often makes a wasted trip to the job site, and someone has to pay for that lost time and money. Your plumber might eat this cost once or twice, but if it becomes routine, you will either lose the plumber or find bills for the lost time in your final billing. I have a clause in my plumbing contract that allows me to charge the person who is contracting my services for any wasted time or trips that result from jobs not being ready. I expect other plumbers do too.

Broken fixtures and missing pieces are not uncommon in the plumbing trade. If you are supplying any of the materials for final plumbing, you must check them carefully. It is probably best to have your plumbing contractor supply all materials for the plumbing work. By doing so, it takes a good deal of responsibility off of you, and it shouldn't cost you much more money.

Avoiding trade wars

Conflicts between the trades can ruin your job. I've seen trade wars where some very serious damage was done on jobs. With such wars, destruction is aimed at one trade, but it often affects several trades, including the general contractor and the property owner. If your trades don't work well together, you've got to get control over them or find new subs. Let me give you just a few quick examples of what trade wars can do to you and your jobs.

Finding the drain-stopping culprit

My plumbing crews were working on a large townhouse project when we experienced an abundance of drain stoppages. This was unusual because new plumbing typically doesn't have this problem. The drains in the six townhouses worked fine at all inspections, but once people moved in, the drains were beginning to clog. All of the stoppages were located on the ground floor, just under the concrete slab. It took quite a while to determine the problem, but we finally figured it out.

After a number of service calls to the townhouses, it became obvious that something under the concrete was seriously wrong. Had pipes been broken? Did the pipes shift and become back-graded? We didn't know until we used jackhammers to break up the new concrete floors. You will never guess what we found. Someone had poured pipe glue down the pipes, probably when they were in the groundworks stage, and then dumped hundreds of roofing nails into the pipes. As the glue set up, the nails were held in place. This created a trap for anything coming down the drain, except clear water, which was used to test the system. Water could flow through the pipes without any problem, but if any type of solid came down the pipe it would result in certain stoppage.

As we moved from townhouse to townhouse, we found the same problem in all six of the units that shared the main foundation. Who would have done such a thing? We didn't know. I heard sometime later that electricians had sabotaged our plumbing. In retaliation, my plumbers, without my permission or approval, got even by pulling the metal staples out of the electrician's rough-in, then reinserting them through the wires. This, of course, created all sorts of electrical shorts and problems that were not discovered until the townhouses were finished. Opening up finished walls to track down the shorts was a very expensive proposition, not to mention an irritating one.

I can't condone what my plumbers did to get even, but I have to admit that they were creative and effective in seeking their revenge. Trade wars are not good for anyone, but they do go on and sometimes roar among people in the same trade. I've seen foundation contractors fight from the tops of their foundation forms right down into the dirt of my groundworks. This activity is, of course, counterproductive, but it happens. As a general contractor, you have to stop it. There is no way you can run jobs efficiently if your subs are fighting amongst themselves.

Locating the leak

The examples I've given you are not isolated events. Here's another. I had done the rough-in on a house one time and had all of the plumbing inspected, which required a pressure test on all piping. During the inspection, all of the piping checked out just fine. When I returned to set fixtures and turn on the water in the basement at the main water valve, however, I heard water rushing above me. At first I thought a hose bib was open, but than I noticed water was pouring through the ceiling of the first floor, right over the formal dining room. Keep in mind, this house was finished. Hardwood floors, which don't like water, were installed and so was wall-to-wall carpeting. When I went up the stairs from the basement and saw water rushing out of the ceiling, I felt sick. Naturally, I rushed back to the basement to cut the water off, but the damage had been done. The ceiling was ruined, the flooring was submerged, and I was in big trouble.

I was puzzled by the leak. Had a soldered joint broken loose after the rough-in inspection? This was the only conclusion that made sense. Well, when I cut open the ceiling, I saw very quickly what had gone wrong. The HVAC contractor had come into this job after me. Apparently, my water pipes were in the path selected for ductwork. I had no way of knowing this when I installed the piping. The HVAC contractor had taken a chain saw to my copper water tubing. There were two half-inch water pipes just hanging in the ceiling. It is certainly no wonder that so much water poured through the ceiling.

Finding what I did in the ceiling was, in a way, good news. It meant that the damage caused was not my fault, but that of the HVAC contractor. My work had been tested, inspected, and approved at the rough-in stage. If the HVAC contractor hadn't hacked the pipes out, this flood would not have happened. The amazing thing is that no one ever brought the cutting of these pipes to my attention. I could have rerouted the piping if I had known they were in the way. But no, some mechanic cut out my pipes, installed ductwork, and never told me about the severed pipes. This story is a long one, but you get the idea. A little cooperation between the trades would have been very helpful in this situation.

Some trade war activity is not destructive to a job, although it is childish and dangerous. For instance, I've seen ladders removed so that roofers couldn't climb down, insulators trapped in attics and smoked out with fire barrels, and a lead carpenter thrown off the roof of a three-story townhouse. The list goes on and on. Trade wars have no place on your jobs. If you have subcontractors who can't get along, replace them with more professional contractors.

Developing teamwork

If you had to describe the best way to get your subcontractors to work at maximum efficiency, the word to use would be teamwork. Having subcontractors who work together for the good of all concerned is the dream of every experienced general contractor. This development of teamwork can be nourished by you, the general contractor, on your jobs.

There are times when a small plumbing crew needs some extra help. For example, if they have to carry a cast-iron tub up a flight of stairs, a few extra hands would be wonderful. Carpenters who are willing to help will make friends for life. Cast-iron bathtubs frequently weigh more than 400 pounds, so all the help a plumber can get is appreciated. Likewise, if your electrician is having trouble snaking a wire through an existing wall and a plumber offers his or her help, the job gets done faster and a bond is developed between the trades. Strive to see these types of relationships formed. They are hundreds of times better than trade wars.

As a general contractor, you are the supreme leader—that is why you are called a general. You have to earn the respect of your subs. Control without respect doesn't mean a lot. Someone who is only afraid of you won't stick around long. Devote your effort toward getting all of your subs working in unison. This will not only make the plumbing phases of your jobs go better, but it will make your entire business more efficient.

4

Decoding code-related issues

What do you know about code-related issues as they pertain to plumbing? If you're like most general contractors, your knowledge of the plumbing code is probably not very extensive. Your knowledge of this code, however, can save you time, trouble, frustration, and money.

The plumbing code is not an easy topic to try to explain. Many experienced plumbers have trouble understanding the code—and they're governed by it. If plumbers can't sort out the code, how can you expect to do it? The code can often be written in hard-to-understand language, but its principles are easy to understand and comprehend.

I used to teach code classes at a technical college. Plumbers would come to my class to prepare for their licensing exams. While some had a good understanding of the code, most of them did not. A few of my students had been journeyman plumbers for years and still couldn't figure out how to size pipe with the guidelines provided in a code book. There were many other aspects of the code that these plumbers had limited or no knowledge of. After a few weeks in my class, many of these plumbers were able to use the plumbing code with good efficiency. If you really want to understand the plumbing code, you should consider taking such a hands-on code class.

This chapter is not as effective as a code class, but it will give you an excellent primer on what the plumbing code can mean to your jobs. There are three major plumbing codes in use. Each of these codes is a little different. To complicate matters, local code officials can amend the code they adopt to suit their local requirements. A customized code varies slightly from the boilerplate code, but the differences can be enough to cause you trouble. The only way to be sure that you are dealing with current, applicable codes is to learn your local code.

I can't predict what code is being used in your area. Even if I knew which of the major codes was being used, I could not be certain of local variations. For this reason, I can only talk with you in general terms. You will have to accept what I give you in this chapter as reference material used to illustrate how plumbing codes work. Again, don't assume that the information is the current plumbing code being used in your region. Before you depend on any code information, check with your local code enforcement officer.

The following are a few important key code issues. My explanations and examples will be based on common code requirements; however, they may not mirror the requirements in your location. Illustrative examples that reflect some of the requirements and differences between the three major plumbing codes can be found at the end of this chapter. Remember, the text and illustrations in this chapter are just examples. Check on local restrictions and requirements.

Interpreting the code

When you read a code book, you might think that what you read is the code; however, there is a major loophole in the code. Although it should be taken verbatim, code officers have a right to interpret the plumbing code and their interpretation might not be the same as yours. This can be more than just confusing, it can be downright disgusting. Because plumbing inspectors can use their own opinion to interpret the meaning of the language in the code, it leaves a plumber or, in this case, a general contractor, wondering what the code actually says.

In the past, I've had plumbing inspectors interpret the plumbing code in some pretty weird ways. Reading the code book on a particular issue seemed to make the rules and regulations very clear, but when an interpretation was handed down by the inspectors, my understanding of the written words was worthless. This hasn't happened to me often, but it has taken place on more than one or two occasions.

Most plumbing inspectors are fair people who don't want to make life harder for plumbers and general contractors. However, if a contractor shows no respect for a plumbing inspector, the odds of him or her having a tough time during an inspection are pretty high. Many inspectors will be happy to help you with questions and problems, as long as you approach them properly. However, if you come at them with a know-it-all attitude, you might just find out first-hand how the power of interpretation can affect your job.

Visiting your local code enforcement office

A friendly visit to your local code enforcement office can do a lot for the future of your jobs. As a general contractor, you may have to deal directly with building inspectors, but you probably don't have a lot of contact with plumbing inspectors. Maybe you should. Inspectors are sometimes a little more forgiving and cooperative with contractors who they know mean to do the right thing. Let me give you an example.

I worked as a plumber in Virginia for seven years with a lot of different plumbing inspectors. I did good work and always intended to install my plumbing in compliance with the code. After a while, to my advantage, the inspectors began to know me pretty well.

I never paid off an inspector to look the other way. And even if I had wanted to, I don't think any of them would have accepted a bribe. Some contractors treat their inspectors to freebies, such as fishing trips and other group outings, but I never felt that this was proper or professional. My relationship with the inspectors was based on pure professionalism and hard work. This honest and professional attitude afforded me some advantages.

One such advantage was the time when I was installing groundworks and had not yet finished the work when an inspector showed up. By rights, the work should have been done and tested before the inspector arrived. Because the inspector knew my work and my integrity, he passed my work before it was done or tested. Technically, this was not right, but I never cheated. I always finished the work properly and tested it. Did the inspectors ever stop back by to see if I did? I don't know. There was a lot of trust and respect between the inspectors and myself, which made my life easier. I never took advantage of the situation, however, by slipping shoddy work in.

I'm not suggesting that you will have inspectors who will approve your jobs before they are finished, but my past experiences were not unique. As a matter of fact, I still have inspectors in Maine who pass my work before it is done because my reputation is good and they can trust me. This trust has been built over time and through my diligent performance in the field.

I have a procedure that has worked well for me over the years, and it might work for you. When I'm getting ready to work in an area where I don't know the inspectors, I make a point of meeting them before I start working. My approach is usually in a manner where I ask their advice. Everyone enjoys feeling important. People like to pat themselves on the back for having answers to questions. If you create a situation where an inspector can enjoy this type of feeling, you're on your way to a winning relationship.

Reading the code book

Code books are not the type of reading material that most people would take with them on vacation—they are not light reading. In fact, many people are intimidated when they thumb through a code book, and I can understand why. If you look at the language and illustrations in this book, they can appear cryptic and difficult to understand. First impressions can be deceiving though. Don't get me wrong, the plumbing code book does have some complicated sections, but most of the code is not as mysterious as you might think.

Code books are broken down into sections so they can more easily be used as a reference. For example, if you are not sure how far apart hangers should be spaced for a horizontal drain pipe, you look up the topic in an index and turn to it. There you will find either text or a table that will give you the information needed to determine hanger spacing. With some exceptions, reading and understanding the instructions is not difficult.

There are three primary plumbing codes in use, each of which requires the use of a specific code book. The three code books are similar in content. As we move through this chapter, I will be talking about the code book used in my area; however, I am familiar with all three. Some of the books use tables and charts to provide certain information while other books use text. For example, I might make reference to a table being used and when you go to look the information up you might find that the same material in your code book is in text form. This will not make a big difference in the use of your book. Now, let's move through some topics for which residential plumbers might use their code books.

Glancing through the administration section

The administration section of the code book deals with general code requirements, such as when you need a permit, who can obtain one, and when inspections are required. There are many other subjects included in an administration section, but as a general contractor, you will probably not need to refer to this section very often.

Reviewing the general regulations

In my region's code book, there is a section on general regulations that deals with such issues as health and safety, pipe protection, toilet facilities for workers, and other information. There is information in this section that can be helpful to a general contractor. Let me give you a few examples.

In the section pertaining to the protection of pipes, a general contractor could benefit from the requirements surrounding various types of soil conditions. If you or your excavation contractor will be digging trenches for your plumbing contractor, you should review the requirements set forth by the code. For instance, my local code requires that rock discovered during trenching be removed to a minimum of 3 inches below the grade line of the trench. Once this is done, sand is required to stabilize the bed of the trench. My local code prohibits pipe and fittings from resting on rock at any point in a trench. This type of information is useful to general contractors who participate in plumbing installations.

If you are about to backfill a plumbing trench, you will find detailed instructions in your plumbing code. For example, my local code requires backfill material to be free of rocks, broken concrete, frozen chunks, and other rubble. The backfill material is supposed to be installed in 6-inch layers. Each layer must be tamped before an additional layer is added. This type of information is needed by the person responsible for backfilling a plumbing trench, be that the plumbing contractor or the general contractor.

The information examples I've just given you are presented in clear, easy-to-read text in my code book. There is nothing mysterious about the information.

Checking for approved materials

The types of materials approved for various plumbing uses are listed in the plumbing code. This section of the code is one that would not normally be used by a general contractor, but you might find it interesting to explore the options for various types of materials. If you want to do this, you shouldn't have any trouble understanding the code requirements.

My code book presents material options in tables. If I want to know what types of materials may be used for a building sewer, all I have to do is refer to the table provided in the code book. In that table I would find eight types of materials for use. My choice would be a schedule-40 plastic pipe, but there are other options, such as cast-iron pipe, available. Again, the information is presented in an easy-to-understand fashion.

Choosing the correct joints or connections

If I need details on joints and connections (Figs. 4-1 through 4-4), I can turn to a section in my code book and find the needed informa-

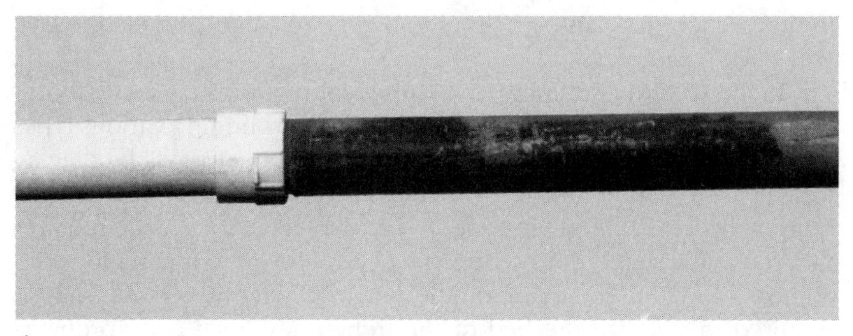

4-1 *A plastic female adapter used to join plastic pipe with steel pipe*

4-2
A coupling band used to make joints with hubless cast-iron pipe

4-3 *A rubber coupling used to join pipes of different types*

4-4 *Two rubber couplings being used to splice plastic pipe between cast-iron pipe*

tion. Again, this section will probably not be of much use to the general contractor, but the plumber will find it very necessary in daily activities. As with the aforementioned sections of my code book, this section is also user-friendly.

Addressing the drainage systems

Drainage systems are addressed in their own section of the plumbing code book. This section should be reviewed by general contractors. There is some pertinent information in the section on sanitary drainage systems that deals with pipe sizing, an issue about which most general contractors will need information. Admittedly, pipe sizing can be confusing, but sizing pipe with the tables provided in my code book is not all that difficult.

For example, let's say that you are interested in sizing building drains and sewers. The only two pieces of information necessary before you can determine an accurate size by using the plumbing code book tables are: what is or will be the grade of the pipe and how many fixture units will discharge into the drainage system. In my code book, I've given four options for pipe grade. The options are laid out horizontally in the table, with a listing of various pipe sizes vertically on the left. To size the drain, a separate table, which is adjacent to the sizing table, gives me fixture-unit ratings for various types of plumbing fixtures. For example, the fixture-unit table tells me that a residential toilet is rated to have four fixture units, a residential shower has two fixture units, and a residential lavatory is rated for one fixture unit.

Using the fixture-unit table, I can count up the number and type of fixtures discharging into a drain to get the total number of fixture units. A small home would probably have around 20 fixture units discharging into a sewer. The exact number would be determined by counting and rating all of the plumbing fixtures in a home.

If I were going to install a sewer with a quarter-inch-per-foot grade, a 3-inch pipe could carry 42 fixture units. I know this because of the data provided in the aforementioned sizing table. Once I know the pipe grade and the number of fixture units being discharged, it is a simple matter to look at that sizing table and see what options are available.

The following is another example of how the drainage system section can help the general contractor. For instance, if I were contracted to put an addition on a house and had to determine the minimum pipe grade requirements for the new bathroom, I could just look in my plumbing code for the table on pipe grade requirements.

The three listed options in that table include using pipes with diameters of 2½ inches or less, installed with a minimum grade of ¼ inch per foot; 3 to 6 inches, installed with a minimum grade of ⅛ inch per foot; and 8 inches or more, installed with a minimum grade of ¹⁄₁₆ inch per foot. Seeing what the minimum grade requirements are will tell me all I need to know to figure out if there is enough height difference between existing plumbing and new plumbing to accommodate a connection without the use of a pump.

Assessing indirect waste piping

Indirect waste piping is not extensively used in residential applications, but it is used. Two examples of an indirect waste pipe are the standpipe for a washing machine (Fig. 4-5) and the air-gap that is created when a dishwasher is installed. A section in the plumbing code deals with indirect and special wastes. As a general contractor, there is not much in this section of the code that will require frequent attention.

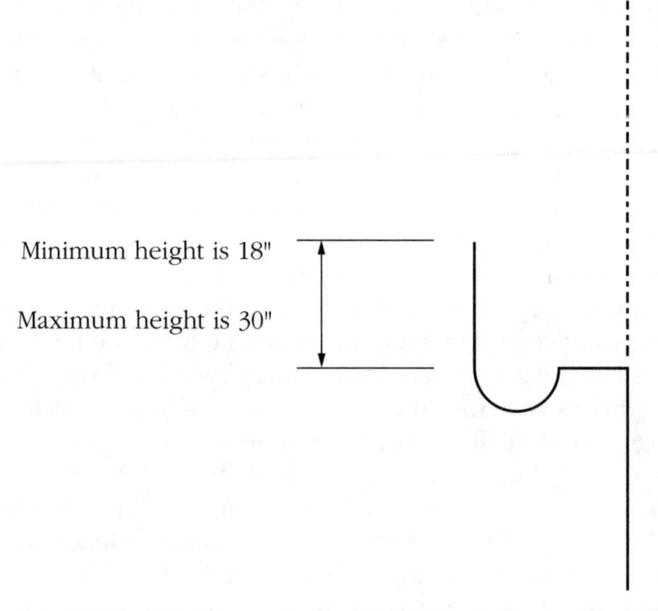

4-5 *Height regulations for washing-machine standpipes*

Researching vents and ventilation

Every home has plumbing vents as part of its plumbing system. Because vents and ventilation are such a complex part of the plumbing systems, general contractors can benefit from learning as much as possible about them from the local plumbing code. The section on venting in my code book is much larger than the section on drainage piping. This would surprise a lot of people.

Sizing vents is a process that is not always simple, but the sizing tables in my code book are easy to understand. For example, let's say you are installing a basement bathroom that will empty into a pump station. The sump that contains the sewer pump must be vented. This is normally done with a 2-inch vent, but how far can the vent run? It can run a lot farther than will be necessary. To determine the exact length, just check the sizing data in the plumbing code. Let me explain.

When I look at a sizing table in my code book to size a vent for a sump, I see vent sizes laid out horizontally across the top of the table. The left side of the table has a vertical listing of the number of gallons the pump will discharge per minute. If I use a 2-inch vent with a pump that pumps up to 40 gallons per minute, there is no limit to the length of the vent. If the discharge rate goes up to 60 gallons per minute, I can extend the vent for no more than 270 feet. Obviously, I would probably never find an occasion to extend a vent for 270 feet in a residential application. As the number of gallons per minute of discharge increase, the allowable distance for extending a vent decreases. The data included in the sizing chart in my code book is easy to understand.

If, as a general contractor, you read and understand the section of the plumbing code dealing with vents, you might avoid some problems on your next job. By studying this part of the plumbing code, you will gain an understanding of where, when, and how vents may be installed. Knowing this will make it easier for you to plan your carpentry work. This could save you time, trouble, and money.

Following the fixture trap rules

A section in the plumbing code dictates rules and regulations for fixture traps. Plumbers must be aware of these requirements, but the information has little effect on general contractors. There should be very little reason for a general contractor to be concerned with this section on traps.

Installing pipe cleanouts

Pipe cleanouts (Figs. 4-6 and 4-7) are required by the plumbing code to allow access to clear drain stoppages. The rules on this issue are not difficult to understand, but they are important. Information in the plumbing code dictates at what minimum intervals cleanouts must be installed. There are also requirements pertaining

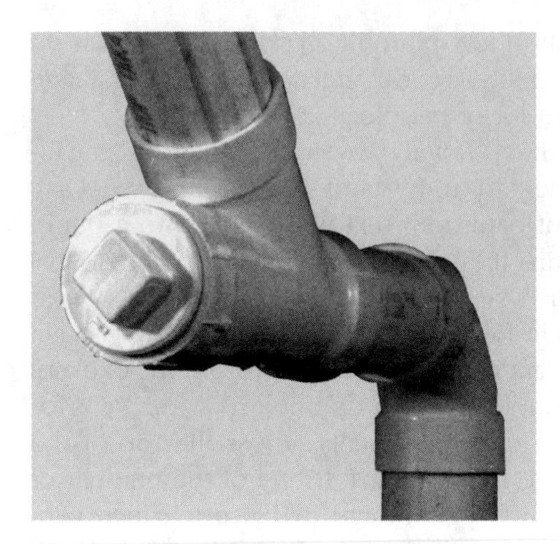

4-6
A cleanout fitting and plug installed in a standard fitting

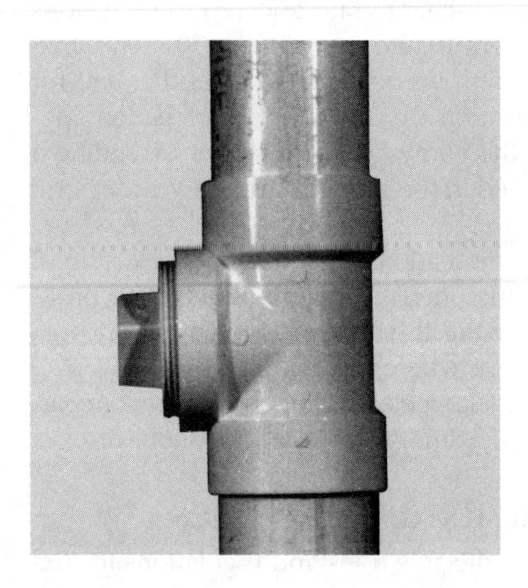

4-7
A test tee fitting being used as a cleanout

to the amount of open space that must be left for access to cleanouts. Open space issues can affect a general contractor. Because general contractors control the other trades such as drywall installers, who might conceal a cleanout, it is important for a general contractor to know what type of access is required. For example, average residential pipes require a minimum of 18 inches of clear space in front of them under the rules of my local code. It is a code violation to conceal a cleanout with a permanent cover. If a cleanout will be concealed in a wall, a removable cover must be used to guarantee access to the cleanout.

Leaving space for fixtures

Plumbing fixtures are covered in depth in the plumbing code. There are rules for what minimum fixtures are required in various situations. You will also find spacing requirements (Fig. 4-8 through 4-14) in this section of the code. These regulations pertain to people who are framing spaces into which plumbing fixtures will be installed. As a general contractor, you are responsible for the framing crew, so you have to make sure they leave enough space for your plumber's fixtures and space requirements.

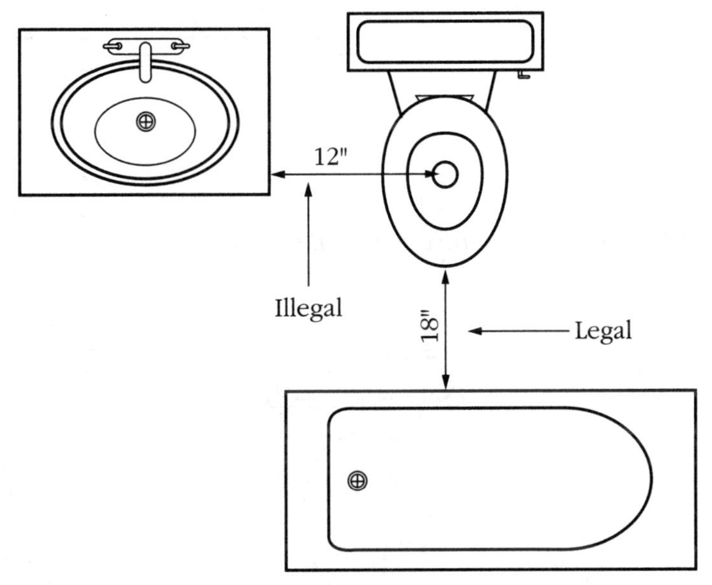

4-8 *Illegal fixture spacing*

Let's talk about fixture spacing in specific terms. If you are framing an area to contain a standard, residential toilet, the compartment must be at least 30 inches wide. A toilet must be allowed 15 inches of clear space on each side of it, measured from the center of the drain pipe. How much open space is needed from the front edge of a toilet to another fixture such as a vanity? The minimum distance is 18 inches. When a lavatory is installed, it must also have 15 inches of clear space on either side of the center of its drain. Being a builder, you need to know this type of information.

When you review the section on fixtures in your plumbing code, you will find a wealth of information pertaining to garbage disposals, dishwashers, and all other types of residential fixtures. You should take some time to read up on fixtures.

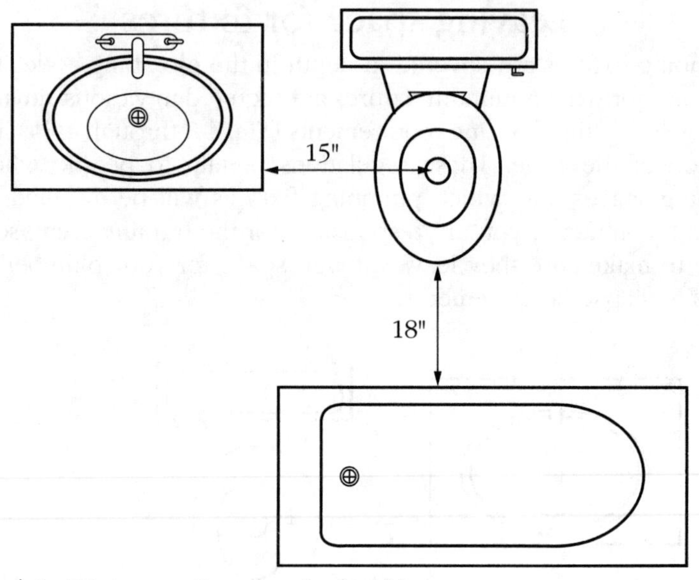

4-9 *Minimum distances for legal layout*

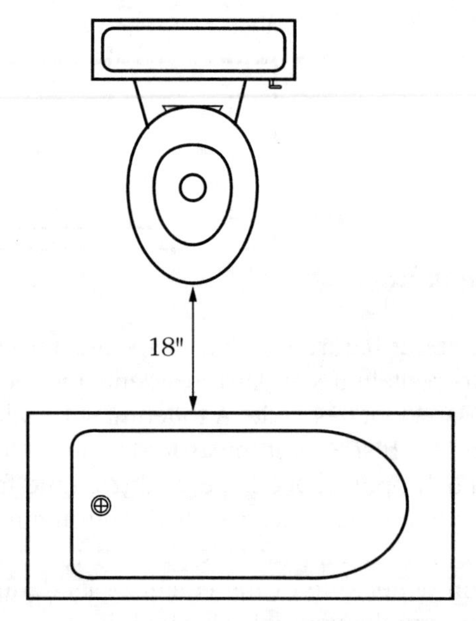

4-10 *Minimum distance in front of a water closet*

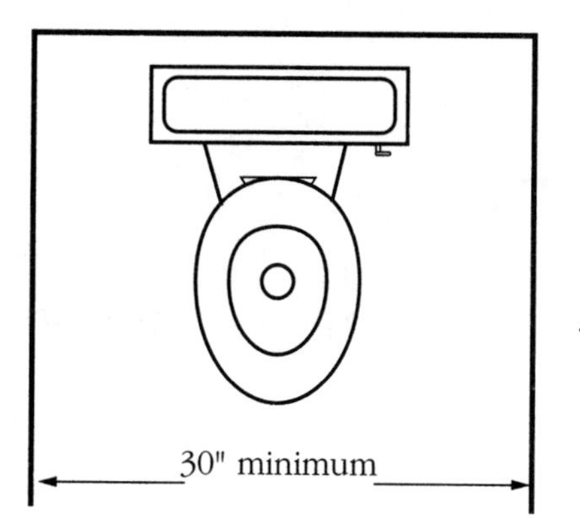

4-11 *Minimum width requirements for a water closet*

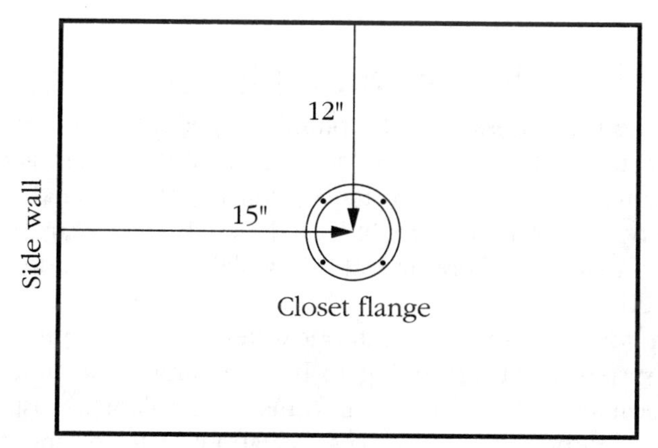

4-12 *Rough-in dimensions for a closet flange*

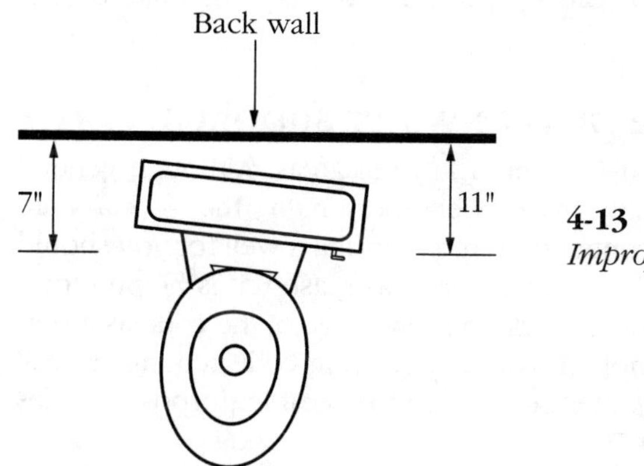

4-13 *Improper toilet alignment*

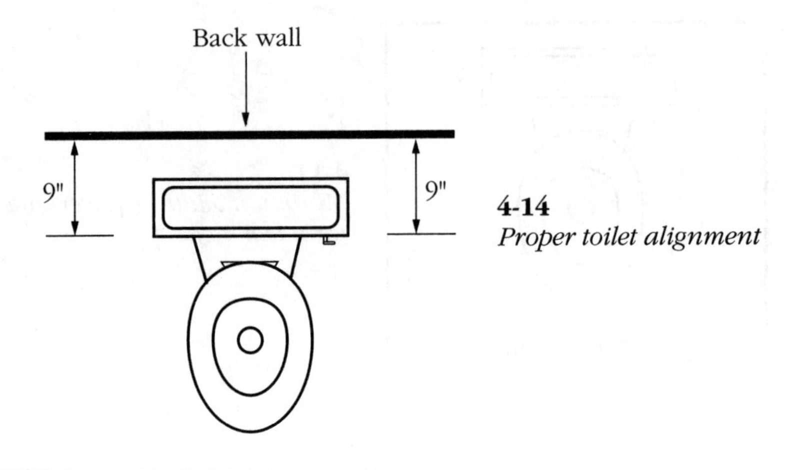

4-14
Proper toilet alignment

Sizing water piping

Water piping is a big issue in the plumbing code, but you, as a general contractor, will not need to know much of this information. Just become familiar with it. For instance, you might not need to know about sizing water pipes, but if you'd like to check up on your plumber in this area, there are tables available in the code book to help you do just that.

When you get into reading about water piping, supply, and distribution systems, you are going to be spending some time on the subject because of the wealth of information available. Most general contractors will not need to be so savvy about water piping, though, as long as a reputable plumber is on the job. If you have time, it will be worth your while to read up on the code requirements for this phase of plumbing.

Installing private water supplies

Private water supplies are common in rural areas. Although general contractors don't normally install wells personally, they do arrange for their installation. A contractor hired to install a well for you should take care of all code-related issues. For this reason, it is helpful for a general contractor to have a working knowledge of the code as it pertains to wells. For example, my local plumbing code requires a well to be made so that the water being used for potable purposes comes from a depth of at least 10 feet.

Other information contained in this section of the code pertains to the distance required between a well and a source of potential contamination, such as a septic tank. Because you may have to plan the locations of wells and septic systems on some jobs, review the requirements in the plumbing code as they pertain to your work and region.

Consulting the code book for other information

The plumbing code contains a lot of other information, such as instructions and requirements for inspections and what to do with storm-water drainage. Depending upon which code is in use in your area, it could take a considerable commitment on your part to learn all of the code requirements. Unless you plan to become a plumber, though, there is really no reason why you should take on the task of absorbing all of the plumbing code. You can pick and choose the information as needed. I firmly believe that you can benefit from acquiring and reviewing a plumbing code book that is being used in your local area. For more information on the plumbing code, please see Appendix A. Reviewing this information now will make the rest of the book more useful.

5

Choosing the right work group

When you address the issue of your plumbing needs, you have to weigh the advantages and disadvantages associated with each work group: employees, piece workers, and subcontractors. These differences are substantial and can mean the difference between profits and poverty. Not all business owners are aware of the various advantages and disadvantages between different types of work groups. If you fall into this category, you will really appreciate this chapter.

Do you know how much your employees actually cost you on an hourly basis? You might be surprised at the number of business owners who have no idea what the true cost is for their employees. Is there really a difference between piece workers and subcontractors? Both of these groups are independent contractors, but there is a significant difference between the classifications. Learning the ins and outs of hiring people is a complicated process. It is also a task that can be quite expensive.

As a general contractor, you will come into contact with numerous jobs involving plumbing. Most generals bring in independent plumbing contractors to handle the plumbing requirements on jobs and then moan about their high cost. Do you believe that your company might show higher profits if you put a plumber on your payroll? Some contractors do believe this, and they are sometimes correct in their assumption. A plumbing crew can even be developed to bring in a lot of extra income. More often than not, however, hiring hourly plumbers as regular employees is a mistake. How will you know, though, if and when to create your own plumbing division?

I've hardly started this chapter, but already there are several thought-provoking questions to be answered. As I'm sure you've guessed by now, the topic of this chapter deals with employees, piece workers, and subcontractors and how to choose the most beneficial plumbing crew for your company. By delving into this topic, we may unleash a wealth of profitable ideas for your business.

One of the safest ways to grow as a business owner is to review historical data. From this information, you can determine which types of actions have worked in the past and which have failed. Creating your own personal historical data from which to pull is an expensive proposition. The trial-and-error effort needed to produce results that can be evaluated will undoubtedly result in financial losses because such a test is not effective until you find some methods that simply don't work at all. If everything you try works, you are cheating yourself by stopping while you're ahead. Until you learn the limits of your business, you may be stopping yourself from huge financial gains. So how can you learn from mistakes without making them? The solution is simple. Learn from the mistakes of others and profit from them. This is sometimes easier said than done, but with the help of this book, you can do it.

Determining differences between work groups

The difference between piece workers and subcontractors is a subtle one. They share many of the same advantages for the people for whom they work. There is, however, one major difference between these two types of workers: Both are independent contractors, but piece workers normally work exclusively for one person, while subcontractors typically work for several different people. The level of control is another big difference between these two work groups. You have more control when you deal with piece workers, but you also have more responsibilities because you must keep them busy. Subcontractors are less easy to control because they don't count on you to keep them busy and often expect to work with a number of contractors to fill their meal ticket.

Contracting piece workers

From a general contractor's perspective, one of the biggest differences between piece workers and subcontractors is that piece workers are almost like employees. They typically work only for one contractor. The good part of this is that piece workers should be there when you need them and be able to take direction a little better than regular subcontractors. The downside to using piece workers is the demand that they put on you to keep them busy. If you don't have a volume of plumbing work at your disposal, they won't stick with you for very long.

Piece workers make their money based on production. If you're a contractor who is not strong in organizational skills and planning, piece workers will not be happy working for you. They simply can't make much money if you don't give them the opportunities they need. You must make the decision to use piece workers based on a variety of factors. For instance, early in my career, I worked as a piece worker for a plumbing contractor. The initial part of the arrangement went very well. I hired a helper to work for me as an employee and made very good money. Because I made so much money, the contractor started mixing my work schedule. This resulted in a decrease in my salary, major frustration, and a move to another company. The actions of the contractor made no sense to me at the time, and they still don't make sense. Let me explain what happened.

I was contracted as a piece worker to do plumbing installation on new houses. The company provided me with a detailed payment schedule for the various types of work I would be doing and established a set price. In the first few months of the arrangement, everything went well, but as I progressed through the deal, things started to deteriorate.

Being a hard-working, aggressive plumber, I turned out a lot of work. It was not unusual for me to install the groundworks for two houses in a single day. I could rough-in a two-and-a-half bath and have it ready for inspection in less then three full days. It was no problem for me to set all the fixtures for these houses in one day. Now I didn't work just 8 hours a day. Working 10 or 12 hours a day was not unusual. My helper and I worked very hard to make a lot of money, and we did.

At the time when this deal was in effect, an average weekly income for plumbers in my area was around $400. I was making an average of $1500. Out of this, I had to pay my helper about $300, which meant I was making about three times what I could have made as an employee. The income made me one happy plumber.

My production as a piece worker was far above that of most other piece workers employed with the company. This should have been a good thing since for every dollar I was making, the company was making money. I can't say what they were making off of my work, but I would guess it was at least equal to what I was making and probably more. Why would anyone tamper with such a good situation? I don't know, but the owners of the company did.

After a month or two of cutting checks for me that were considerably higher than other piece workers, the company owners decided to lower my income. This, of course, cut their own throats. If I accomplished less to make less, they also made less. Nevertheless, my work schedule was altered. All of a sudden, I was having to travel to two or three jobs a day, instead of going to one job and remaining there to achieve maximum production. In the process, I was losing billable time by traveling and many of the jobs were not even ready for me when I got there. This resulted in more lost time and money. It didn't take much of this to drop my income down to around $1000 a week, a loss of about $500 a week, because the owners were jerking me around.

I sucked up the losses for a few weeks, thinking of the good of the company. When it became obvious to me that the owners were moving me around on purpose, I got angry. When this happened, I terminated my contract and went to work for one of their competitors as a piece worker. The new company didn't throttle me down. Consequently, I stayed there until I went into business for myself as a full-scale plumbing contractor.

When you use piece workers, you want the fastest, best workers you can get. They will make a lot of money for both themselves and for you. If you use piece workers, don't make the same mistake that my first company made with me. That company lost a very productive plumber and provided their competitor with a tool to use against them. You can't afford to make this type of mistake.

In order to keep piece workers happy, you have to keep them busy. Because they only make money when they produce completed jobs, you have to create enough work to keep them going. You also need to arrange material deliveries and schedule work for the piece workers in such a way that does not create downtime. Piece workers who are riding to a supply house to pick up parts or standing around while carpenters make framing corrections are losing money. Good piece workers won't take much of this before they move on to another contractor who will keep them busy.

Employing subcontractors

Subcontractors offer general contractors many of the same advantages associated with piece workers, but with one added benefit—very few subcontractors work for a single general contractor. Most subs typically work for a number of general contractors and for individuals who are not professional contractors. Because they are more independent than piece workers and you are not their only source of income, subcontractors might not respond to your requests with the same level of intensity as a piece worker. Conversely, the advantage is that you are not expected to keep subcontractors busy all the time. Almost any work you throw their way will be appreciated, and most of them will not abandon you when work is hard to come by.

Both piece workers and subcontractors are independent contractors. As their work provider, you are not obligated to the same responsibilities you would have with employees. Subcontractors and piece workers are cost-effective alternatives to employees in many, but not all, cases. If you need one focal point to separate subs from piece workers, it should be that piece workers are dedicated primarily to you and subcontractors have many employers.

Hiring employees

It is very difficult to choose the right work group to complete your work. Previously, I discussed piece workers and subcontractors along with their advantages and disadvantages. Although similar in some respects, hiring an employee can be a much bigger responsibility for you. Your biggest advantage to hiring employees is control, because you are their only source of income. This is a decisive advantage, but you can achieve nearly as much control with piece workers while avoiding the costly aspects of employees.

During the nearly 15 years that I've been in business for myself, I've used employees, piece workers, and subcontractors. There have been both good and bad points associated with each type of worker. Overall, I rate employees as being the lowest on my scale of desirability. The many reasons for choosing employees last are addressed in the following section.

Having control of your workers

Having control of your workers is key to getting a job done right and on schedule. The amount of control you actually have varies with the different work groups. For instance, the amount of control you have over employees is typically superior to the control you might have over piece workers and subcontractors. Some of this is just in the nature of the different types of work groups. When you deal with independent contractors, you are not allowed to control them completely. This is what makes them independent contractors. If you want to dictate all aspects of what goes on during a normal work day, set detailed work rules, and enforce those rules, you should hire employees, not independent contractors.

When you hire employees, you can set forth specific hours of the day for which they must be available for work. You can't do this with independent contractors. It is permissible to establish acceptable work hours, but you can't demand that independent contractors work a set schedule. For example, you could say that all work must be done from 8 a.m. to 6 p.m., but you can't insist that independents be on the job for all of these hours each day. Your deal with independent contractors is based on production, not the hours worked. If you assume too much control over independents, your insurance company or the Internal Revenue Service will deem them to be employees. This defeats the whole purpose for using independents.

Being neither a lawyer nor a tax expert, I'm not qualified to delineate the differences between employees and subcontractors. You should talk to qualified professionals to see what you can and can't do without triggering the status of employees when you think you have independent contractors. In general terms, you have much more control over employees than you do with independent contractors.

Determining the cost of each work group

The cost factors separating employees from piece workers are extreme. You may pay a much higher wage to piece workers, based on an average hourly earning, than you would to an employee. This doesn't mean that the piece workers are costing you more than employees. In fact, the opposite is more likely to be true. Most contractors who use piece workers effectively find them less expensive than employees. The same can be said for subcontractors in a large number of circumstances.

Employees are an expensive luxury for a business owner. Their cost goes well beyond the hourly wage you agree to in an interview. To illustrate this, let's look at some of the extended costs of hiring employees to do your plumbing. The first cost is the agreed-upon hourly wage. This is an obvious expense. For the sake of our discussion, let's say the hourly wage for your journeyman plumber is $12 an hour (this number is just a benchmark to work from).

The type of plumber you hire also makes a difference in the total hourly cost of the labor. For example, a service plumber can cost more per hour than a plumber who does only new construction work. Why is this? The expense of buying and stocking a service vehicle is much more than the cost of sending a plumber to a specific job site in the plumber's personal vehicle. Plumbers who work construction jobs don't have to be outfitted with company trucks that carry thousands of dollars in stock. The same might be true for some types of remodeling jobs, but most work of this nature requires a plumber to have a stocked vehicle on the job. For the sake of our discussion, we will assume that you are providing a vehicle, major tools, and some basic stock for each plumber you hire. If you're looking to set up a service division, the projected costs will be higher than what is discussed here. Conversely, if you are going to have your plumbers report to a stocked job with their hand tools each day in their own vehicles, you cost will be lower. Keep in mind that all figures used here are hypothetical.

If you hire your own plumbers at an hourly rate of $12, you know the wages are your first expense. But how much will the overhead associated with the plumbers be? This is a complicated issue, and it will vary with each individual plumber and company. For the sake of this example, I will use figures that are realistic in my opinion but may not work out in reality for your situation.

The plumbers' trucks will be the first overhead expense. Let's say that your truck payments for each vehicle are $275 a month. This figure doesn't include the down payment that was made on the vehicle, a cost that should be factored into the overall cost of your plumbers. In an attempt to keep this example easy to understand, though, we will leave down-payment expense out of the equation. In addition to the monthly payment on the truck, you have to factor in the cost of insurance. We will say that the insurance is $50 a month, bringing the revised monthly vehicular expense up to $325, excluding operating cost. When you add in gasoline and maintenance, your total is up to $600 a month.

How many billable hours will your plumber turn in during the course of a month? New construction plumbers should average at least seven billable hours a day. The rate for service plumbers could be less. In our example, we will assume a billable time of 35 hours a week. I want to stress, however, that 24 to 30 hours of billable time is not unusual in the plumbing business. Assuming that 35 hours is the billable rate and that you give your plumber two weeks of vacation time each year and pay $600 a month for a vehicle, you are now looking at an increase of what the plumber costs you of about a little more than $4 an hour. In other words, your $12-an-hour plumber is costing you $16 an hour, and we are far from done in the assessment of total cost. In this projection, we have made no adjustment of the money you must spend to stock the truck, an amount that could reach $10,000 or more for a service plumber. Construction and remodeling plumbers can get by with much less inventory, but the cost of everything on the truck should be accounted for in your overhead projections.

Stocking the truck

The cost of truck stock is difficult to translate into an hourly expense. Plumbers who are doing new construction work don't need a lot of stock. Service plumbers, however, do require extensive stock to be efficient in their duties. If you were to buy a new truck and set it up for a service plumber, the cost of tools and stock could easily exceed $10,000, and this doesn't count any of the expense of the truck itself. The $10,000 could be spent only on company-provided tools and rolling stock. Because the stock should turn over and make money for your company, it is hard to say what the real cost of this expense is when based on an hourly overhead. One way or another, however, you have to account for the cost in your overhead projections and budget. This, by the way, is an expense that would not necessarily have to be incurred with piece workers or subcontractors.

Allotting for vacation

Employees today expect employer-paid vacation time. Many plumbing companies limit paid vacations to a one-week duration, but a lot of companies authorize two weeks. You're in control of this decision, but you must make your company benefits competitive with other plumbing outfits, or you'll have a hard time keeping quality plumbers on your payroll. Paid vacations are easier to put into perspective on an annual budget.

Let's say that your company policy allows plumbers two weeks of paid vacation each year. To calculate this cost effectively, you must know the true hourly expense of your individual employees. Your hourly cost will be considerably higher than the wage at which your plumbers were hired. Overhead expenses drive the hourly costs up a great deal. To keep our math simple for this example, let's use the hourly rate of $12 to pinpoint the cost of a two-week paid vacation. Again, keep in mind that the true cost will be much higher. Eighty hours of paid time, at $12 an hour, is worth $960.

In our cost-evaluation example, we are saying that your plumbers bill out 35 hours a week. Given a two-week vacation, this number will be based on a 50-week year. Of course, our numbers are very conservative in this example. Actual time billed out will probably be much less, but to remain consistent in our example, we will stick with 50 weeks at 35 hours a week. When we divide out the cost of a vacation, which is $960, by the number of billable hours a plumber works, you have an added hourly cost of 55 cents an hour. Based on this, your new hourly cost is $16.55.

Calculating in sick days

Many companies allow their employees some number of paid sick days each year. This is time when you are paying your plumbers even though they are not generating income. If we say that you allow six sick days a year, this adds $576 a year to your overhead. This is assuming that your employee uses all of the sick days and that the rate of pay is $12 an hour. Converting this cost to an hourly overhead gives you a number of 33 cents an hour. Now your cost per hour is up to $16.88.

Offering insurance benefits

Insurance, such as health, dental, disability, and life, is often given as a company benefit. Most employees pay some portion of these costs, but the employer often pays the lion's share. The amount of money expended on an employee's insurance by a company can range drastically in its amount. For the sake of our example, let's say that your company participation adds up to $350 a month or $4200 a year. Divided out, your participation in insurance expenses adds $2.40 to the hourly cost of your employee. This brings your hourly cost up to $19.28. Your $12-an-hour plumber is starting to get expensive!

Factoring in the other costs

Before you can conclude your total hourly cost for each plumber, you must factor in other costs, such as liability insurance, worker's compensation insurance, holiday bonuses, and so forth. Before it is all over, the actual cost of your $12-an-hour plumber may end up being $24 an hour.

The cost of most plumbers is not double the amount of their hourly wages, but the figure is not normally too far from it. In terms of cost projections, you should be safe if you double the hourly wage to arrive at the true cost, although this formula doesn't always hold true. For example, overhead expenses can be less if you have two plumbers working out of one truck or if you give one week of paid vacation instead of two. To arrive at a true cost for each of your employees, you have to do the math using numbers that apply specifically to your company. These examples should, however, be helpful to you in pinpointing the actual cost of your employees. There is no fair way to compare the cost of piece workers, subcontractors, and employees until you know the overall cost of your employees.

Employing piece workers

Piece workers can be a very cost-effective alternative to employees. As independent contractors, piece workers don't expect the same benefits as employees. For example, paid vacations and insurance benefits are not normally extended to piece workers, and most piece workers provide their own transportation and tools. Your savings from using piece workers can amount to several thousand dollars a year. Not only that, you are not faced with all the paperwork that goes along with employees. When you choose to use piece workers, your paperwork is limited to contracts and a 1099 form at the end of the year. The reduction in paperwork eliminates administrative costs that must be applied to the total hourly cost of employees.

Piece workers are production-oriented, meaning they get paid for work they properly complete. The more work they do, the more money they make. This is a big difference when comparing them to employees, who typically get the same paycheck each week regardless of their production. If employees rough-in two houses, they get paid their standard rate, not more just for roughing-in another house. Likewise, roughing-in only one house doesn't cause the employee to lose pay. The bottom line is this, employees are not as motivated to achieve results as piece workers.

The potential problem with the production-oriented attitude of piece workers is that they may cut some corners along the way and lower the quality of their work. Having work meet minimum code requirements is usually the basis on which piece workers operate. If you are dealing with customers who want their plumbing rough-in installations to be pretty as well as functional, piece workers could present you with some problems. This, of course, is not true of all piece workers. Many independent contractors take just as much, if not more, pride in their work than employees. However, because employees are being paid by the hour to do exactly the type of work you require, they can afford to take their time and turn out beautiful rough-ins, a luxury most piece workers can't afford.

If you are going to demand superior quality from piece workers, you should make this point very clear when you are negotiating payment rates and terms. If they did the work in accordance with local code requirements and the contract between the two of you, there is no room for you to fuss when a customer complains that soldered joints weren't wiped down. Think about this type of situation before you engage piece workers. By laying out a standard for them to work by when they work for you, you may be able to avoid confrontations later.

Hiring piece workers for new work

Piece workers can be hired to do remodeling, service, and new work, but they are best suited for the latter. When you are discussing prices with piece workers, it is very much like dealing with subcontractors. Technically, piece workers are subcontractors. This is why they don't add to your company's overhead expenses like employees do.

Hiring piece workers for new work is simple. You provide the piece workers with your company guidelines that explain what you expect of them in terms of professionalism. A set of plans and specifications for the jobs being awarded are provided to the independent contractors. A firm price is agreed upon and contracted for. This is sometimes done on a job-by-job basis, but it is also done on a generic basis. Let me explain this a little further.

When I worked new construction plumbing as a piece worker, I worked exclusively for one company at a time. The first company I dealt with made a deal with me on a generic basis. A fee schedule was set up for different types of plumbing work. The fees floated with the amount of fixtures involved. For example, I was paid a certain amount for each toilet being installed, so much for each lavatory installed, and so forth. This allowed the plumbing company to bid work with confidence. They knew exactly what their labor cost was going to be for any job based on the fixtures being installed.

From a contractor's perspective, generic pricing is very attractive. It is not as good for a plumber who is a piece worker, but the arrangement evens itself out most of the time. Let me give you an example. If I were given a three-bathroom house to rough-in this week, it might require a lot of drilling to install the plumbing and its groundworks might have to be installed in hard clay. All of this would slow me down, and I wouldn't make as much money as I would like. However, the house given to me the following week might not require much drilling at all and the groundworks might be installed in sand. The speed of doing this job would produce earnings that were higher than expected. When the two jobs were averaged out, I would have done okay.

As a contractor who does new work, you can set up your piece workers in one of two ways. You can have them bid every job or you can set flat-fee rates on a per-fixture basis. It is less time-consuming for you to bid work when you have flat-fee prices with which to work. Having your workers bid each job is effective, but it takes longer and by the end of a year, there is probably not going to be a lot of difference in the amount of money made regardless of the way you reach your price. The decision of which path to take is, of course, yours. I prefer the flat-fee rate from a contractor's point of view and the bid-each-job method from a plumber's position.

Estimating the cost of remodeling work

Remodeling work is never predictable. If you plan to use piece workers to plumb remodeling jobs, you almost have to work with a bid-each-job method. It simply isn't practical to set generic rates for this type of work. I suppose it could be done, but I've never known a contractor to use a flat-rate plumbing fee for remodeling work.

Setting the rate for service work

Setting generic fees for service work can get complicated, but it can be done. In fact, many contractors apply generic fees to what they will pay piece workers who do service work. For example, a piece worker might be paid a flat fee of some amount to clear a stoppage in drains with diameters of 2 inches or less and a higher fee for drains that are larger than 2 inches in diameter and no more than 4 inches in diameter. Similarly, a set fee can be assigned to the replacement of a kitchen faucet, the replacement of a toilet, or the installation of a garbage disposal. There are many ways to use flat fees with service plumbers.

Some contractors set up their piece workers who are doing service calls on a percentage-of-the-take basis. If a service plumber collects $75 in labor and material markup on a job, the company splits the gross take with the plumber. The percentage of the split varies from company to company, depending upon the amount that you and your plumbers agree to.

I've consulted with a number of small plumbing companies who were convinced that there was no way they could use piece workers in their service and repair divisions. The simple methods that I've just described work very well. Piece workers can be used in any aspect of the plumbing trade. Some adjustments are needed in procedures, depending upon the type of work being done, but piece workers can replace employees in all aspects of field work. The amount of money saved by taking this approach can be considerable, so the idea is worthy of your attention.

Using subcontractors

The use of subcontractors is normally the best course of action for general contractors who need plumbing work done. There are many advantages to working with subcontractors and few disadvantages. To elaborate on this, let's look at both the pros and cons, so that you can come to a comfortable conclusion on how to have your plumbing requirements met.

Advantages of hiring subcontractors

There are many advantages of working with subcontractors. Some of the biggest advantages include not having to pay for worker's compensation, health, life, dental, and disability insurances; not having to pay a percentage of their Social Security fees, as you do with employees; and having a set cost for your job because everything is based on a contracted flat fee.

Another advantage is that subcontractors provide their own transportation. This saves you from buying trucks and rolling stock. Most subcontractors provide all of their own materials for work being done, which alleviates one more headache. Paying employees to stand around waiting for work to come in can cripple your company. This is a fear that is eliminated when you use subcontractors. You only pay subcontractors when there is work available.

Callbacks can kill a small company. If you have payroll plumbers who create circumstances for a lot of unpaid warranty work, your profits drop quickly. This is not the case with subcontractors. Plumbers you retain as subcontractors are responsible for correcting their own mistakes, at no cost to you. This can be a very big advantage.

A failed inspection can cost you several hours of paid time when you're using employees. This doesn't happen when the work is contracted to independent subcontractors. Assuming that your contract stipulates that all work done must meet or exceed local code requirements, additional work required to bring an installation into compliance will have to be done by the subcontractor, at no additional cost to you.

Liability insurance isn't a back-breaking expense, but it is a substantial one. When you require your subcontractors to insure themselves, and you, for mistakes they make, you eliminate yet another overhead expense. All of these savings add up quickly to save you money. Of course, you are paying subcontractors more than you would employees, or so it would appear, but the overall cost of subcontractors at the end of the year may be much less than you expect.

Tools are another issue to be considered. Outfitting payroll plumbers with all the tools necessary to do their jobs can be very expensive. The maintenance and repairs of these tools can also add up. Then there is the theft insurance that should be carried to replace stolen tools. All in all, the cost of tools for a large plumbing company can run into tens of thousands of dollars. This is another expense that can be avoided when you use subcontractors.

I could go on and on with the advantages of subcontractors. When you start to break the comparison between subcontractors and employees down into small pieces, subcontractors are way out in front. Unless you have a constant flow of plumbing work to do, subcontractors or piece workers are your best bet. Whichever type of independent contractor you choose to use, you can cut your plumbing overhead considerably.

Disadvantages of hiring subcontractors

The disadvantages of using subcontractors are not numerous, but they do exist. Control over the workers is probably the biggest factor to take into consideration. When you have employees, you have more control over their actions than you would with subcontractors. If you like being able to insist that plumbers follow your instructions to a tee, you probably will not be happy with subcontractors. Because piece workers and subcontractors are both independent contractors, there is only so much that you can do to control them.

The cost of using subcontractors is usually more on a per-job basis than what you would expect from employees. Most subs charge a higher hourly rate than what you would expect to pay a payroll plumber, but the final cost of a job might be less when subcontractors are used. Before you can make this determination, you must know the true cost of your employees. If you factor in all of the expenses of maintaining an in-house plumbing crew over the course of a year, subcontractors may turn out to be a bargain.

With my past experience, I can come up with only two disadvantages to using subcontractors: lack of control and money. There are, however, ways to work around these problems if you have a volume of plumbing work. You can hire one plumbing crew as employees and use subcontractors or piece workers for the work that this crew can't handle. By having one plumbing crew on staff, you can defuse many of the problems encountered with subcontractors.

It is possible to use your one payroll crew to respond to jobs where subcontractors are dragging their feet. When subcontractors and piece workers make a mistake, you can make a deal with them to allow your hourly crew to correct the deficiencies. This allows you to charge time for your hourly crew to the independents, but it also allows the faster workers (i.e., the subs and piece workers) to keep pounding out production. Under a high volume of plumbing work, it is not cost-effective to pull production people off of their current jobs to go back and correct minor code violations and leaks. This is an ideal job for your hourly people, and most subs and piece workers will welcome the opportunity to keep up their production by paying your staff to fix their little glitches.

Making your hiring decision

You must make the decision on whether to hire subcontractors, piece workers, or employees, and it will not be an easy one. Most general contractors do better when they use subcontractors. Builders who have large projects to work on do very well with both piece workers and employees, although I've always preferred piece workers. Unless you do large jobs or are setting up a complete plumbing division that will serve the public as well as tending to your company needs, I would recommend that you avoid hiring employees.

It takes a lot of plumbing work to keep good plumbers busy throughout a year. Building 12 houses a year may seem like a lot of work as a builder, but it is only a drop in the bucket for a good plumbing crew. If you happen to have a townhouse project to build where hundreds of units will be turned out each year, you should be able to make more money by hiring plumbers as employees or piece workers. The same might be true if you have substantial commercial jobs, such as schools or hospitals, to build. On the whole, however, it doesn't pay for general contractors to set up their own plumbing divisions to do in-house work. Again, if you will market your crews to other builders and homeowners, chances are an employee plumbing crew might be to your advantage. Weigh your decision carefully and don't make it in haste.

6

Purchasing the tools of the trade

If you're thinking of setting up your own plumbing division, you need to consider purchasing the tools of the trade. Depending upon the scope of your plumbing endeavors, this can be a very expensive proposition. Specialty tools, such as electric pipe threaders and jackhammers, don't come cheap. Many of these tools can be rented if they are needed only occasionally, but sometimes it is best just to buy them. To determine what to buy and when to buy it, you have to evaluate your total plumbing needs.

Most plumbers expect to provide their own hand tools. By hand tools, I am talking about tools that fit into tool boxes and that don't require anything more than physical force to operate. Power tools and more expensive specialty tools are generally provided to plumbers by the companies for which they work. Some of these tools include electric drain cleaners, pipe threaders, jackhammers, and a host of smaller tools, such as pipe thawers.

The types of tools needed by plumbers depend heavily on the type of work being done. A service plumber needs tools that a construction plumber would almost never use. Plumbers who specialize in remodeling work require tools that neither construction nor service plumbers have much need for. With this in mind, you have to identify the type of plumbing work that your company will be doing before you can set up a plan for the purchase of tools. It is also wise to consider renting expensive tools until you establish a level of need for them within your organization.

In order to give you a clear overview of the tools that plumbers use, I think it best that we discuss the primary tools on an individual basis. In some cases, such as with hand tools, I will lump a lot of tools into one category, but most of the time I will separate and describe them individually. I will also describe the types of work for which each type of tool is normally needed. For example, a jackhammer is almost never needed in new construction plumbing, but it can see frequent use by plumbers engaged in remodeling.

Right-angle drill

A right-angle drill is an absolute necessity for all plumbers. Although service plumbers don't use this tool as often as plumbers who are installing new systems, all plumbers depend on this tool. This drill has a half-inch chuck, and a variety of drill bits can and should be purchased for it.

Special kits of plumber's bits are on the market. These kits contain drill bits in the most common sizes used by plumbers. Self-feeding bits for plumbers are not cheap. A bit used for 2-inch pipe can cost around $50 at wholesale prices. Bits for 3-inch pipe, which don't come in the kits, can cost around $100. The complete set of needed bits could easily set you back $350 or more. This is more than the drill costs!

Reciprocating saws

Reciprocating saws see nearly daily use by plumbers engaged in new work, service work, and remodeling jobs. Fortunately, there are not a lot of expensive accessory items needed to complement a reciprocating saw. Cheap versions of these saws start at around $100, if you catch them on sale. A dependable, professional model will cost closer to $200.

Hand tools

Once you have a good reciprocating saw and right-angle drill, the next items to consider are hand tools. Most professional plumbers already own their own hand tools and should be willing to use them as a part of their job. If you have to fill a tool box for a plumber, be prepared to spend several hundred dollars. A huge list of tools can be thrown into the category of hand tools. Some of them are mandatory, while many of them are used only occasionally, but if you need a tool and don't have it, you've got a problem.

Some hand tools are needed by all types of plumbers. For example, a basin wrench (Fig. 6-1) is needed when installing and replacing sink and lavatory faucets. Adjustable wrenches, socket sets, hammers, offset pliers, screwdrivers, and similar assorted tools are used by all types of plumbers. A grade level (Fig. 6-2), however, will almost never be needed by a service plumber, but anyone installing new plumbing can benefit from this tool. Grade levels are a type of specialty tool that

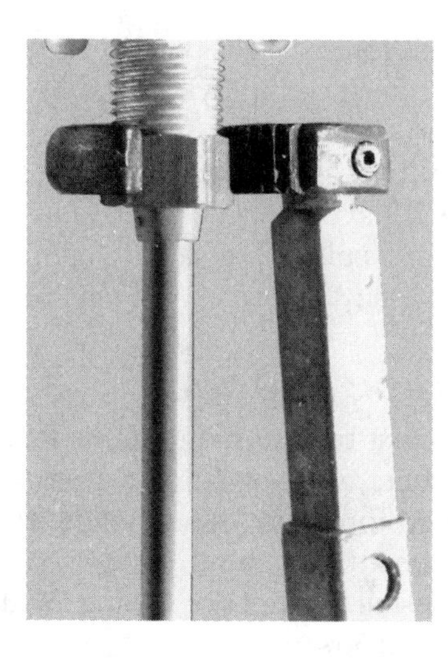

6-1 *Basin wrench*

6-2 *Grade level*

allows plumbers to install piping on a set grade or pitch with ease. A 6-foot level would probably never be taken out of a service truck, but plumbers installing new bathtubs and showers will use such a tool frequently.

Hand tools must be tailored to the type of plumbing work being done. A handle-puller, used to remove handles from old faucets, is not needed in new construction, but it is a valuable tool for a service plumber. Once you establish the type of plumbing your crews will be doing, you can select the tools needed by talking to your experienced crewmembers. Keep in mind, though, that most plumbing companies require their plumbers to provide their own hand tools. This fact should save you some money.

Pipe cutters

Pipe cutters can be tiny hand-held devices or big, heavy tools, depending on the type of pipe being cut. Cutters for copper tubing are considered hand tools, and therefore, should be part of the package any professional plumber brings to the employment table. Larger cutters, like those used for steel and cast-iron pipe, are normally supplied by the plumbing contractor. In this case, that means you. It's entirely possible, however, that you will have no need for large cutters. The two types of pipe that require big cutters are steel, which is used for gas piping, and cast-iron, which may be used for drains or vents.

Steel pipe cutters

Steel pipe cutters are needed most by plumbing outfits that do work with gas piping. If your plumbers will not be doing gas work, you probably won't have much need for heavy, steel cutters. However, if the buildings you plumb are served by natural gas or propane, typically the plumbing crew will install the interior piping. With this being the case, you will need the cutters.

Cast-iron pipe cutters

Cast-iron pipe is not used nearly as often in modern plumbing as before, but there is still a lot in use and being installed. For example, it is not uncommon for commercial buildings to have cast-iron pipe installed for underground drains, and many high-class residential builders use cast-iron stacks in the homes they build to deaden the sound of water draining down the pipes. Plumbers who do remodeling work come into frequent contact with cast-iron pipe. If your company will be working with cast-iron, you should purchase cast-iron cutters. There are two basic types to consider: snap cutters and ratchet cutters. The type of work you're doing will dictate which ones you should buy.

Snap cutters

Snap cutters are used to cut cast-iron pipe when new installations are being made. These cutters resemble a very large pair of scissors with its blades fitted with a special chain of cutting wheels. These cutters are used best when the pipe being cut has not been installed yet and can be cut while lying on the ground or floor. Snap cutters are not a good choice for plumbers involved in remodeling work, but they are fast and effective when cast-iron pipe is being used for new installations.

Ratchet cutters

Ratchet cutters use the same type of chain cutting wheels as the snap cutters, but they don't look or operate like snap cutters. The result of the two types of cutters is the same, but the mechanics are very different. Ratchet cutters have only one handle, while snap cutters have two. A lot of space is needed to use snap cutters, but this is not the case with ratchet cutters. Remodeling plumbers can benefit greatly from a good set of ratchet cutters because they can be used in confined conditions.

If you will be installing cast-iron pipe in new installations, opt for snap cutters. When your work will be in the fields of repair and remodeling, ratchet cutters will be the tool of choice. These cutters are not inexpensive. You can expect to pay upwards of $300 for a set of them, but they're worth every penny when they are needed.

Pipe-threading equipment

Pipe-threading equipment comes in two versions: manual and electric. Manual versions are acceptable for occasional use, but electric models are best when a large volume of pipe needs to be threaded. When we talk about pipe threading, we are usually dealing with gas piping, which is made of steel. I have already discussed the pipe cutters for this work, but the cutters are the cheap part of the deal, even though they are far from inexpensive. In addition to cutters, a plumber needs a vise, dies, an oiling can, and possibly some other tools. Making a commitment to go with electric threading equipment is quite expensive. We're talking thousands of dollars. Manual equipment is much cheaper and just as good, but slower.

Unless you are in an area where most construction contains gas piping, setting up a complete threading operation with electric equipment will take a long time to pay for itself. This is certainly one time when renting may be your best way around an expensive proposition. Using manual equipment is a viable alternative, but you lose some time and money in labor costs. The following will provide more information so that you can make an informed decision on what type of threader to purchase.

Power threaders

Power threaders are nice, but they cost a bundle of money. Unless your plumbing division is involved in a lot of work with steel pipe, buying an electric threader is probably not a good idea. If you happen to get a big job where a lot of steel pipe threading is needed, you can rent a power threader; otherwise, make do with a manual threading procedure. Your plumbers won't appreciate the extra physical labor, but your bank account will reflect your savings.

Back when galvanized steel pipe was used for drains and vents, threading pipe was a big part of the plumbing trade. Because this practice is all but extinct, the only pipe threading done by most plumbers is for gas pipe, which is black steel. In areas where plumbers also install hot-water heating systems, some pipe threading is done for pipes connected to boilers and oil tanks, but nipples are often used to eliminate the time spent threading pipe. Therefore, if your company will not be doing extensive gas piping, you probably don't have a need for a power threader.

Manual pipe threaders

Manual pipe threaders will build biceps quickly. Threading pipe the old-fashioned way is pure physical work, which can be invigorating, but it can also send you to the whirlpool with sore muscles. As a contractor, the benefit of using manual threading equipment is the cost—arm-powered threaders cost a fraction of what electric threading machines cost.

When should manual threaders be used? They are very efficient on pipes with small diameters. If your people are working with ½-inch or ¾-inch pipes, a manual threader is not too physically demanding and works well. Once the pipe sizing reaches a diameter of 1 inch, 1½ inches, or 2 inches, manual threading takes on a new level of pain, especially in the larger diameters. Any pipe over 2 inches in diameter should be done with an automated system.

When I entered the plumbing trade, I was classified as a helper. This meant that I drilled holes, ran jackhammers, and threaded pipe, among other undesirable tasks. I vividly remember threading 2-inch steel pipe by hand for two plumbers at a time as we installed gas piping in a commercial building. It was a very hot September in Virginia, and I was standing on black pavement that seemed to be melting the soles of my boots as I cranked away on the handle of a ratchet threader. Cutting for two plumbers and advanced helpers meant that I was in almost constant motion. Previously, I thought drilling the steel beams for hangers was a bad job, but I soon discovered that threading the pipe was worse. The only good thing to come of this time of my life was learning to be a good plumber and losing a lot of weight. If you had asked me back then, I would have traded almost anything I had for an electric threader. But, as a contractor, I can see why a manual threader is beneficial.

When you plan to use manual threading equipment, there are several things you will need: a simple die set, for pipe up to 1 inch in diameter; a ratchet-type tool, for larger diameters; a sturdy vise, typically a tri-stand chain vise; and cutters. All threading equipment requires you to oil the threads regularly to avoid damaging your ties.

Automatic oilers are available, but an oil can that is used manually will also work. Reamers are recommended for clearing fragments from pipes. When you get into threading pipe, you have to consider a number of tools and expenses.

Electric drain-cleaning equipment

Electric drain cleaners are almost mandatory for service plumbers, but they are seldom used by remodeling and new construction plumbing. If your plumbing work is centered around remodeling and building, the occasions when you will need electric drain-cleaning equipment will not warrant its purchase. You will be much better off to rent such a machine on the rare occasions when it is needed.

Jackhammers

As a building contractor, you may already own a jackhammer. If your plumbing division is doing predominately new work, there will be few occasions when they will have any need for this tool. However, if you are into remodeling, there might be a number of times when a jackhammer will serve the needs of your plumbers well. The best example of this is related to the installation of bathrooms in basements. This type of plumbing installation requires concrete floors to be broken up, and a jackhammer is the tool of choice for this job.

During my career as a plumber and builder, I've found electric jackhammers work well. Air-powered jackhammers have more power, but they are inconvenient and overpowered for most of the residential work I do. If you need a jackhammer at all for your plumbing division, a good electric one should be adequate. These tools can be rented at almost any full-service tool rental center. I purchased my own jackhammers because of the many basement bathrooms my company has installed, but this doesn't mean that you should invest in one also. I always recommend that people rent specialty tools until they establish an ongoing need for them. Once you see that you have enough business to justify the purchase of a tool, you should buy it. Prior to this determination, just rent it.

Miscellaneous tools

Miscellaneous tools for the plumbing trade can cover a lot of ground. You could consider purchasing: a bullet heater, to thaw pipes in the winter or to keep your crews warm; an electric pipe thawer, if you have a service-oriented plumbing company; air compressors, for testing new installations; or a generator, if you have a new plumbing division. If you put your mind to it, you can come up with a very long list of potential tools and equipment to buy for your new plumbing group. Most of these purchases should be deferred until you prove the worth of your plumbing outfit. For all you know at this point, the venture will be a flop. If it is, you're left with a bunch of plumbing stuff that you have little to no use for. This is not something that you want to have happen.

I could probably write an entire book on the subject of tools used by plumbers. It would be a bit boring, but there is no doubt that there are enough tools in use to fill a book. This doesn't mean that you have to or should buy all of them. Your goal as a profit-oriented contractor is to equip your plumbing division in a way that is profitable without being oppressive. Spending several thousand dollars for tools might give you a tax write-off, but it doesn't guarantee your success in the plumbing business. It may turn out that you will be better off to sub your plumbing work out to independent contractors. If this is true, you try to determine this before you make major capital investments in tools, equipment, and inventory.

Surveying your crew

One way to determine what you need to buy is to talk to your crew. Be advised, however, that most working plumbers want every tool imaginable to do their job, but the majority of their requests, although preferable, will not be needed. You have to make the ultimate decision on what will be purchased.

I've given you a good overview of the types of tools most plumbing divisions require. There are, of course, other tools that I've not detailed. If you are an average contractor, the tools mentioned in this chapter will get you well on your way into the plumbing trade. The master plumber you hire will be able to refine the list. It is up to you to use good judgment in what is and isn't needed. My best advice to you, in regards to high-dollar items, is to rent them before you buy them. You will pay a little more here and there with rentals, but the long-term effect can be in your favor. Once you establish an ongoing need, purchase the equipment.

7

Using specialists or a full-service company

Should your plumbing contractor be your heating contractor? This can be an interesting question. There are pros and cons to using one contractor for multiple trades. It is, of course, easier when you have to deal with only one subcontractor for two phases of work required in your jobs, plumbing and heating, but not all plumbing contractors are great heating contractors. Some plumbers don't work with heating systems at all. There are a number of companies, however, that do offer both plumbing and heating services. In fact, some mechanical companies offer plumbing, heating, air-conditioning, and electrical work on their menus.

As a general contractor, you have to weigh both the advantages and disadvantages of putting so many of your mechanical eggs in one basket. If you run across a superior contractor who can meet all of your mechanical needs, the business marriage between the two of you should be a happy one. However, if the contractor you select is not geared up to meet all of your needs in the most effective ways possible, the relationship might end in an early divorce. This issue clearly requires some consideration on your part.

Taking the easy way out

Having one contractor to meet both your plumbing and heating needs would seem to be the easy way out. Obviously, the fewer people you have to deal with as subcontractors, the easier your life should be. In theory, this is a correct assumption, but, unfortunately, reality may prove otherwise. Taking the easy way out might make your life more difficult than ever before.

Not all plumbing contractors offer service in other fields, such as heating. Some that do shouldn't. It is not uncommon to find plumbing companies where only specialized areas of plumbing service are offered. You might use one plumbing contractor for all of your commercial installations, another company for your residential installations, and yet another for your remodeling work. There are so many possibilities that making a wise decision can require extensive thought.

Using specialists

There is a lot to be said in favor of using specialists. While it can be a bit of a bother to use three different plumbing contractors for various types of work, the rewards may be worth the trouble. For example, plumbers who concentrate all their efforts on remodeling work should be able to produce better work, faster, than plumbers who are accustomed to just doing new installations. The same type of line can be drawn between plumbers who work on commercial work most of the time and those who specialize in residential installations. When a person does one type of work on a daily basis, he or she should be better at the task than someone who spends only a fraction of the time performing similar duties.

Plumbing is a trade where there is a lot of opportunity for specialization. Many plumbing companies do only service and repair work. For instance, there are companies where the main objective is the installation, maintenance, and repair of back-flow preventers, while some plumbing companies spend all of their time clearing drain stoppages. Finding plumbers who specialize in institutional plumbing is also possible. When you consider how many selected fields of pure plumbing there are, it may make you wonder why a plumber would expand out into heating work.

Some types of heating systems lend themselves very well to being installed by plumbers, such as hot-water heating systems. The piping used with a forced hot-water heating system provides installation work that any plumber should be able to perform, but what might be difficult for the plumber are the motors and controls found on a hot-water boiler. Many plumbers are not qualified to work with these aspects of a heating system.

The various components of a hot-water heating system can be fairly complicated. Installing the parts is often easier than troubleshooting and repairing them. If you happen to hire a plumber who has limited knowledge of a heating system, you may get an acceptable installation but have trouble when warranty work is needed for adjustments and repairs. This problem should be addressed before repairs are needed. Having a customer call you in the middle of a cold, winter night complaining of not having any heat will force you to deal with a service issue. If your plumber can't, or won't, resolve such problems quickly, you're going to have some very cold and angry customers beating down your door.

There is another potential problem with using your plumber to install some types of hot-water heating systems. If the system is one that uses fuel oil, your plumber may not be licensed to work on the burner. If this is the case, you have two subcontractors involved in one phase of work. The plumber sets the boiler and installs the piping and baseboard heating units. Then a licensed oil-burner mechanic sets up the burner and fires it. When a problem pops up with warranty work, you have to determine if the person needed for service is the plumber or the oil-burner mechanic. Having this type of separation can really create confusion and trouble when you need service in a hurry.

Just as plumbers can choose which area of their trade they wish to specialize in, so can heating mechanics. A company that installs heat pumps might not be willing to install a hot-water heating system. Depending upon the types of heating systems most common in your area, you might wind up working with several different contractors for your plumbing and heating needs.

Although a lot of plumbing companies do offer some type of heating service, my experience has shown that most limit their heating work to hot-water systems. The first plumbing company I ever worked for provided full-service plumbing, heating, and electrical services. This was more of a mechanical company than just a plumbing company. Finding companies like where I started is not difficult. They can provide great service at good prices, while allowing you to deal with one company for all of your plumbing, heating, air-conditioning, and electrical needs.

Should you hire specialists for your jobs? I think it is a good idea to hire the best people you can find. This may mean having three plumbing contractors, two heating contractors, and so forth. Like other contractors, I prefer to work with as few subcontractors as possible to get a job done. However, I am willing to use a large number of subs if I feel it is warranted to achieve the best possible outcome.

With my business, I make decisions based on experience and gut instinct. Because I am a master plumber, there are not many occasions when I need an outside plumbing contractor. There have been times, however, when I've hired competitive plumbing companies to work for my construction division. Sometimes this was done to get work rolling until my own plumbing crews could get caught up on large work loads. There have been times when I've contracted outside plumbing contractors to install well pumps or to run gas piping in commercial buildings. My decision for doing this was not based on a lack of knowledge in the particular areas of work. I brought in the outside help because they were better equipped to handle the work.

My company doesn't do a lot of well work, so I don't have a truck set up with a boom to pull submersible pumps out of wells or to roll piping off as a pump is installed 400 feet deep in a well. Neither do I concentrate heavily on commercial work. Therefore, I don't own a lot of staging and other equipment that is needed when installing plumbing and gas piping in big commercial buildings. Hiring outside plumbers who do have this equipment and the specialized skills required for a job makes sense to me as a building contractor. I could rent equipment and use my own plumbing company to do the work, but it is easier and faster to bring in specialists.

Think about what I've just told you. I own a plumbing company, yet there are times when I contract competitive plumbing companies to handle my work. If I find this to be cost-effective, your use of specialists should also be worth serious consideration. When people have the right skills and equipment to do a certain type of work, they are the ones to call. Let me give you a quick example.

I told you that my plumbing company doesn't do a lot of work with well pumps. Submersible well pumps are a lucrative business, but it is a part of the plumbing trade that I've never gone after with a vengeance. My company stays busy enough with remodeling, new installations, and routine service work. Because I don't do a lot of pump work, I don't have a truck set up for this type of work, but a friend of mine does. He can go out with his truck alone and pull a defective pump faster than my crews could do it with two people on the job. The difference is in the truck. The pump specialist can also install a pump in a deep well much faster alone than my two-person crew. Again, the key is the truck.

If you've never pulled a pump, you might not be aware of the amount of effort it takes. A pump that is hanging 300 feet down in a well, such as with my personal pump, can be quite a load to pull up. Having a truck with a boom winch on it makes pulling the pump up easy, but the job would be far from easy if done by hand.

Installing a submersible pump in a deep well can also be difficult without a pump truck. Where do you lay out 300 feet of pipe while preparing the pump setup to be installed? If space is limited, this can be a big problem. When I installed my personal pump, I had 25 acres of land to stretch the pipe out over, but not all jobs have this type of land area. Having a truck that is equipped to use large reels of piping solves the space problem. When I built my house, I installed my own pump to save money. If the job was one being done for a customer, I probably would have hired my friend to install the pump. My crew might have taken over at the well casing, but the pump and drop-pipe would have been installed by the man with the truck. This example doesn't involve heating systems, but it is one way of showing you how equipment makes a difference when working with specialists. Now let's look at a heating example.

Hot-water heating systems require the use of a boiler. The weight of a boiler can be considerable. Transporting a cast-iron boiler from a driveway to a building is a job that could require either several strong people or a power tailgate. Couple this power tailgate truck with a strong appliance dolly, and you further reduce the need for people power. If you're paying a contractor based on the number of people on your job, the contractor using manpower may cost more than the specialist who uses mechanical equipment.

Putting heat pumps on the roofs of commercial buildings can be tricky business. A crane is often used for this type of work. In some cases, other types of lifting equipment replace the need for a crane. Either way, whoever is responsible for getting the units on the roof must be prepared for the job at hand. Using human strength and a block-and-tackle arrangement can work, but it is not as efficient and more dangerous.

When you are thinking about using specialists, you must look deeper than the hourly rate. Consider the overall cost of the job. This involves more than just the cost of setting a heating unit. Time is your best friend, or enemy, depending upon how it is used and billed out. If doing a job the old-fashioned way saves you money on one side, it may cost you money in the long run. Don't allow yourself to suffer from tunnel vision when comparing the cost of specialists to those of average contractors. Turn the picture around several times to see it in all its various forms. You might discover, as I often have, that paying more for specialists is not as expensive as it first looks.

Hiring full-service companies

Full-service companies are available and can be a good value as far as cost and convenience. I mentioned earlier that my first plumbing job was with a mechanical company. If you want to reduce the number of subcontractors you have to deal with, mechanical contractors are a sensible route to take. These companies are often quite capable of handling all of your plumbing, heating, air-conditioning, and electrical needs. Most of the companies offering all of these services have their own specialists: a master plumber oversees the plumbing work, a master electrician takes care of the electrical division, and master heating mechanics are in charge of the heating work. Specialists cover every aspect of the job.

Problems sometimes arise, however, when a company offers too many types of work to its customers. It is not unusual to hire companies where the plumbing work done is excellent, but the heating work is only mediocre. The reverse is also quite likely to be true. The plumbing may stink while the heating work is great. When evaluating a multi-service company, you have to investigate each division of the company. If you're the type of contractor who requires references from your subs, make sure you get references from the company for the types of work they will be doing. I've had bad experiences with multi-service contractors in the past, but I've also enjoyed some good working relationships with this type of company. You have to make a decision on a multi-service company in the same way you would for a single-service company, but you have more to check out when dealing with full-service companies.

There are distinct advantages and disadvantages to using one company for all of your mechanical needs. In the following section, I discuss some of those pros and cons.

Using a single company for multiple tasks

One of the best aspects of using a full-service company for multiple tasks is that you are dealing with only one principal. That means that the people in the various trades within the one company are working together, so cooperation between the trades and scheduling should be easier. This is no small gain for a general contractor, because scheduling and cooperation between the trades account for many of the problems that general contractors face.

What are some of the other advantages to using your plumbing company to install your heating systems? Once you have established a solid working relationship with a subcontractor, it is easier to maintain continuity in your jobs. When one contractor is doing both your plumbing and heating work, the desire to have jobs turned out in a consistent manner is more likely to be achieved. Payment for work done might also be easier to manage when using one contractor for two or more phases of work. If you are giving a contractor a lot of work, the contractor should be willing to work with you when construction draws are delayed. This can help your cash flow and lessen the tension on your mind.

Warranty work is another issue where using one contractor for both plumbing and heating could be to your advantage. If you're dealing with hot-water heating systems, it will be much easier to only have to call one company to solve a problem. When a plumber installs part of a hot-water heating system and an oil-burner mechanic does the rest of the work, through two independent companies, placing responsibility for warranty work can be difficult. You will save yourself much time and trouble by dealing with only one subcontractor, provided that contractor is capable of doing both burner work as well as the plumbing.

Dealing with the problems

Problems can come up when you rely on one contractor to serve both your plumbing and heating needs. Any one of these problems can be all it takes for you to decide against using one company for work in two different trades. However, if you research the full-service contractors before you hire them, you can eliminate many of the problems that may arise. Let me give you some examples from my past to illustrate this point.

I used to know the owner of a plumbing and heating company who offered complete plumbing and heating services. The heat work was limited to hot-water heating systems and gas-fired furnaces. Heat pumps were not installed by the company, neither were straight air-conditioning units. The plumbing contractor did, however, have arrangements with an HVAC company to do joint ventures when heat pumps and air-conditioning work was contracted. On the surface, this company looked good, but when you looked a little deeper, there were many problems associated with its organization.

The owner of the company was a master plumber. He possessed a fair knowledge of plumbing, but was not as knowledgeable as some of the plumbers working for him. This in itself could have created some problems. The company maintained two offices, which were located about 30 miles apart. Each office was run by a supervisor. In the case of this particular company, both of the supervisors were licensed oil-burner mechanics. One of the men also held a journeyman plumber's license.

So, we have a company where the owner is a master plumber and the field supervisors are both predominately heating specialists. This is a wise setup for the owner because if one of the heating specialists quit, another was available to pull permits and supervise work. Because the owner held a master plumber's license, he was most vulnerable in the heating side of the business. I can't fault his reasoning in hiring two master heating mechanics.

The employees in the business were young. Many of them had less than a couple of years of experience. Due to the inexperience of the field crews, the supervisors had a large burden on their shoulders. To top it all off, neither of the supervisors had an extensive background in plumbing, and that's the area where the company did most of its work. The owner could have avoided some of the problems by taking a more active role with the employees, but he left the management of his crews up to the supervisors.

The company soon found itself in financial trouble and hired me to fix the many problems. Much of the company's trouble was related to poor management, but some of the problems could be traced to the supervisors and their lack of experience in plumbing. With both supervisors having the most experience in heating work, obviously the heating side of the business ran smoother than the plumbing side. If the company had been structured with one master plumber to supervise all plumbers and one master heating mechanic to supervise all heating work, the problems might not have existed. Because the only master plumber in the entire company, before I became associated with it, was out selling jobs and not supervising field work, the plumbing division suffered. Trucks were stocked with a lot of unnecessary material and not enough material that was needed. Installations were done to code, but barely. A host of problems cropped up due to inadequate knowledge on the part of the supervisors. It is not fair to blame the supervisors, because they were clearly heat experts, not plumbing professionals. This is the type of trouble that you can encounter when you choose one company for both your plumbing and heating needs.

Deciding what is best

Is one company best for providing both plumbing and heating services? This is a question that must be answered on a local and individual level. There are many good companies in business who can provide quality service in both heating and plumbing. If you can locate these companies and work with them effectively, your work load can be reduced. Picking a poor company to do business with, however, will only increase the amount of work thrown upon you. This same rule applies to selecting any subcontractor.

There is no doubt that many companies are capable of providing great service in both the plumbing and heating trades, but whether or not these companies exist in your area is another story. You've seen my opinions of the pros and cons for using one company to perform work in two trades. To complete this discussion, I will leave you with my personal opinions on the subjects.

I favor the use of specialists. This is why I use one framing crew, one siding crew, one roofing crew, and one interior-trim crew when building a structure. There are plenty of companies available who offer all of these services under one roof and I've used such companies in the past. My experience has proved, however, that the results on my jobs are better when I select individual crews for specific types of work. It is, in many ways, simpler to use just one carpentry contractor for all carpentry work, but when I've weighed all of the considerations, I've found that specialists serve my needs best. This may not be true in your case.

I personally don't like the idea of mixing plumbing with heating. I prefer to separate the two trades on my jobs. This is not to say that I never combine the two trades, but that decision is made based on each job as it comes up. At times using a full-service company makes more sense than bringing in specialized crews. The bottom line is you have to match your subcontractors to the job at hand.

If I were asked to make a definitive conclusion on whether or not plumbing companies should be used for heating service, I would say that they shouldn't. This isn't fair to the great number of good combination companies doing business, but my personal results have shown that success is better achieved when the work is done by separate companies. Regardless of whether you use a full-service company or specialists, you have to make sure that any subcontractor you retain is competent to do the work requested. You could very well find combination companies who fit this profile. If you do, use them. By dealing with only one subcontractor, your administrative time and expense will be reduced.

8

Installing the underground plumbing

Underground plumbing, when it is used, can be considered the foundation for a plumbing system. If the plumbing installed underground is not properly placed, the rest of the plumbing system will not work as planned. Many plumbers refer to underground plumbing as groundworks because it is either intalled on top of, or buried in, the ground. In this early phase of plumbing, the ground may not be the base; instead, sand, fill dirt, gravel, or crushed stone may be what supports and surrounds underground plumbing.

When and how the groundworks are installed for a job is crucial to the overall success of a project. This pertains to more than just the plumbing system. When the groundworks are not installed properly, problems can affect concrete contractors, carpenters, plumbers, and of course, general contractors, who are affected when anything on a job goes wrong.

As a general contractor, how responsible are you for making sure that the plumber has properly positioned and installed the groundworks? You can take an attitude that if there is a problem, it's the plumber's fault, but your overall job will be compromised if a serious problem exists. Some general contractors incorrectly assume that local plumbing inspectors will catch and reject problems with underground plumbing. To some extent this is true, but it is not reasonable to expect plumbing inspectors to look for, or find, all of the potential problems that underground plumbing can create for a general contractor.

Plumbing inspectors are looking for code violations. For example, the inspector is looking to see if a plumber installed a drain underground with a diameter of 1½ inches, an obvious code violation. The plumbing code requires that all underground drains have a minimum diameter of 2 inches. However, if a pipe is not properly lined up with the soon-to-be-built partition wall, the plumbing inspector will probably neither notice the problem nor care about it. Getting pipes to line up with future walls is a problem for the plumber and the general contractor. And, I should add, this costly problem occurs often.

Too many general contractors bask in a false sense of security when they know plumbing work will be inspected by a code enforcement officer. It is fair to expect the inspectors to stop inferior work, but you can't assume that they will solve all of your potential plumbing problems. It would be nice if all plumbing companies used some form of quality control, but this doesn't often happen. As a matter of fact, many plumbing companies are unwilling to pay field supervisors to go around behind plumbing crews to check every aspect of their work. It's easy for little mistakes to slip past the attention of a busy field supervisor and become bigger problems. This is when workers and contractors often begin pointing fingers at other people.

Who's at fault? Which person will assume financial responsibility for the problem? These are only two of the questions that are likely to come up when the groundworks for a building are not properly installed. If you don't want to find yourself caught up in such a mess, you have to take on some responsibility for checking the underground plumbing installation. Because you're not a plumber, this can be a tedious task, but it doesn't have to be. The following sections will help you develop the skills needed to keep your plumbers in line and your pipes in walls.

Putting in the groundworks

Underground plumbing, or the groundworks, is the first stage of an interior plumbing job, when it is required. The underground plumbing is the piping that is installed below grade, but within a building. When a concrete floor will be poured in a building, groundworks (Fig. 8-1) may or may not be needed. The determining factors will be the height at which the building drain will exit the foundation and whether or not any plumbing will be installed on the concrete floor. In some cases, there is no need for underground plumbing, within the context of what we are discussing.

8-1
Example of groundwork plumbing

Outside water services and sewers are not a part of this discussion and will not be considered a primary element of groundwork plumbing. It's true that these pipes are normally installed in the ground by plumbers, but, for the purposes of this discussion, groundworks are comprised of plumbing that is located within a building's foundation and not extending more than 5 feet from the exterior foundation wall.

Underground plumbing is needed when you are constructing a building where the sewer or water service will enter the building at a point lower than the lowest floor. When this type of situation occurs, plumbing at this point is rendered inaccessible. This is a common situation in both residential and commercial buildings.

Aside from digging trenches, the installation of underground plumbing is not particularly difficult. There are no holes to drill, and once the trenches are opened up, the work goes quickly. As simple as this work looks, the positioning of pipes in an underground system is crucial. A tiny mistake in this phase of your job can result in a big headache. I've seen plenty of them and know the effect they can have on both the plumbing and general contractor.

Working with make-believe walls

When plumbers install underground plumbing, they often have to work with make-believe walls, or walls made of string, instead of having physical walls in place (Fig. 8-2). The location of these to-be-built walls is determined by blueprints. A margin of error is always present. If a mistake in the placement is made, the result can affect not only the plumbing contractor, but also the carpenters, drywall contractors, general contractors, and any number of other people.

As a general contractor, you should check the final placement of all turn-ups for underground plumbing before you accept the job as being finished (Figs. 8-3 and 8-4). Technically, this shouldn't be your responsibility. The plumbing contractor should assume all responsi-

8-2
The use of strings to indicate wall location

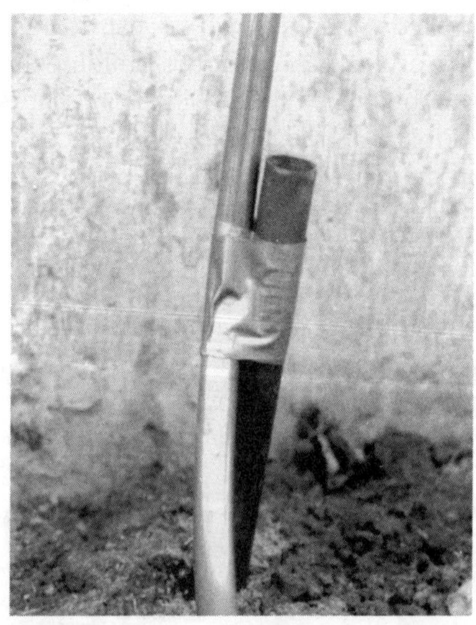

8-3
A water pipe turn-up

8-4
A drain turn-up

bility for proper placement, but if something is miscalculated, that doesn't give you back the time it takes to have the work redone. More importantly, it doesn't make your dealings with an angry customer any easier. The only way you can be sure to avoid problems with the turn-ups for underground plumbing is to confirm their locations yourself.

Checking the position of pipes when they are first installed is important, but it is not the end of the job. A lot can happen to the piping in underground plumbing systems between the time pipes are installed and concrete is poured. For example, the person who is responsible for filling and grading the foundation hole may accidentally shift the pipes. This can happen quickly and easily and the result is just as bad as if the pipes had been installed in the wrong location in the beginning.

Turn-ups in underground plumbing should be checked just before they are surrounded by concrete. Even this is no guarantee that everything will work out well. The force of the concrete being poured can move pipes and it doesn't take much movement to create a big problem. One way to ensure that turn-ups don't move is to secure them either by burying them in dirt, sand, or gravel, or staking them in place with ½-inch copper tubing (Figs. 8-5 and 8-6). Type "M" copper tubing works very well for this purpose.

When a drain or vent pipe is laying on the ground, it can be shifted with very little effort, unless it is secured. A plumber can take a short piece of copper tubing and bend it to create a triangular shape.

8-5
Copper stakes used to maintain pipe location of drain

8-6
Copper stake used to tape a water pipe turn-up to

The tubing fits over a drain or vent pipe. The point of the triangle can be hit with a hammer to drive the legs into the ground. As the apex of the triangle comes into contact with the drain or vent pipe, it holds the pipe securely in place. Over the years, I've tried numerous ways of keeping groundworks where they belong, and I've found copper tubing to be the most effective means of accomplishing this goal.

Pipes can also be secured with ½-inch copper tubing. Instead of bending the tubing and using it in a triangular shape, you can drive straight lengths of the tubing into the ground to maintain a drain or vent position. With a straight stake on each side of a horizontal pipe, it is unlikely that the pipe will be moved enough to cause problems.

When you inspect the work of your plumbers, you should confirm the location of every pipe that is turned up. This will require a second set of hands in most cases, and you may have to pull your own string lines. If you assume that the strings installed by the plumbers are correct, you could be sadly mistaken. Not everyone can read blueprints well and some people are just plain sloppy when it comes to pulling strings to simulate wall locations. Leave nothing to chance. Check each pipe with your own calculations to make sure they are accurately placed. Then make certain that the pipes are secure enough to withstand whatever force might cause them to move. Even with all of this work on your part, it is still likely that some turn-ups will not be where you want them once the concrete is poured. Unfortunately, this is just unavoidable.

Finding solid ground

Groundworks must be installed on solid ground. This is an aspect of a job that any good plumbing inspector will check; unfortunately, not all inspectors take the time to see if the support is solid. While you might think that a plumber would be sure that the pipes being installed will not sink, this is not always the case. We live in a world where the only person you can trust completely is yourself. This is unfortunate, but true. When you delegate duties to in-house supervisors and subcontractors, there is room for error. It is always possible that the best of us will make a mistake, but it seems that some people make a lot more than others.

If you are concerned that your groundworks may sink, causing drain stoppages and possibly broken joints, there is a simple test that you can conduct. Once underground plumbing is installed and turned over to you for payment, go out and walk on it. If the horizontal pipes bend or sink under your weight, the ground under the pipe is not compacted to a satisfactory level. This is not only a workmanship issue, it is a code issue. If you have pipes that move much when they are stepped on, you need to have your plumber rectify the situation before any concrete is poured.

Setting the pitch

All plumbing drains are required to have some amount of downward pitch. This pitch is often referred to as grade. The most common grade used on drains is based on the pipe dropping ¼ of an inch for each foot it travels. Under this scenario, the top end of a 4-foot section of pipe would be one inch lower at the discharge end that it would be at the fixture end.

The amount of fall, grade, or pitch placed on a drain is important because if it is not enough, the pipe will not drain properly and stoppages may occur. For example, if a toilet is flushed into a pipe with excessive grade, the water will rush out too quickly, leaving toilet tissue and other solids behind. The solids can accumulate until the pipe is blocked. To work properly, drains must have a consistent grade that is in compliance with the local plumbing code.

Theoretically, plumbing inspectors are supposed to check the grade of each drain pipe, but with their tight schedules, many times they simply don't have the time to check every one for proper pitch. If you want to make sure that you will not be breaking up a concrete floor to correct drain stoppages in pipes with improper grading, you should check the grade yourself, before the floor is poured. This may sound like a lot of work, but it isn't. If you have a grade level, you can check the grade of drains and vents quickly. This is an investment of your time that can result in far fewer problems in the future.

Installing sleeves

Sleeves are required for the protection of pipes that penetrate foundation walls (Fig. 8-7). It is also necessary to sleeve pipes that pass under a foundation footing. The best time to install these sleeves is before concrete is poured. Installing sleeves after a wall or footing has been poured is certainly possible, but the amount of work required is far more than if they were put in earlier in the construction process.

Some types of pipes, such as copper piping, may suffer from chemical reactions when placed into contact with concrete. When you know you will be using such piping, some form of protection is required. Foam insulation is often used as a sleeve to protect vulnerable pipes from direct contact with concrete (Fig. 8-8). This is, again, something for which code officers should inspect, but it never hurts to check on such items for yourself.

8-7 *A sleeve in a concrete wall*

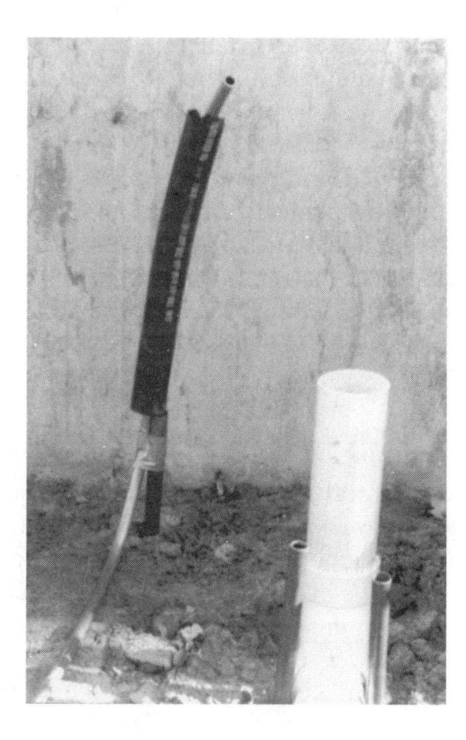

8-8
Foam insulation used to protect a copper water pipe from concrete that will surround it

Sealing the pipes

When groundworks are installed, it is important to seal the pipes used as turn-ups. Leaving pipes open invites unwanted problems. For example, children might find it great fun tossing rocks down the open pipes. Any foreign objects that are placed in the open ends of turn-ups can force you to break up the concrete floor covering the pipe to make repairs. This type of problem would fall under the responsibility of your plumbing contractor, but you would still have to put up with the aggravation. The installation of caps on all pipes can eliminate this risk.

Doing the final grade

The final grade of a finished concrete floor can be crucial in the installation of groundworks. This factor doesn't affect pipes turned up to serve as stacks and vents, but it plays a vital role in the installation of floor drains, closet flanges, and other fittings that are set to coincide with the finished floor.

Most plumbers will require you, the general contractor, to provide a benchmark for the finished grade. Once this is done, the plumbers can go about their work. Minor mistakes in determining the finished grade can result in major problems. If the grade level set during the installation of groundworks is not accurate, pipes might come up out of the floor or be installed so close to the top of the concrete that their presence will cause the concrete to crack. Floor drains might be roughed-in too low or too high. The same problem might affect the installation of toilets and showers. When you establish a finish-grade level for your plumbers, make sure it is accurate and that the concrete crews maintain it. It is also advisable to have your plumber use protective caps to keep concrete from being poured too closely to drains for water closets, where flanges will have to be installed after the floor is poured (Fig. 8-9).

Installing tub boxes

When bathtubs will be installed on a concrete floor, tub boxes are installed during the rough-in of groundworks (Fig. 8-10). Tub boxes are simple, rectangular frames that are intended to keep concrete away from the drain for a fixture. If a tub box is filled in by the flooring crew, your plumber will have a very difficult time connecting a drain to the bathtub. Your plumber should install tub boxes as needed, but you may need to make sure that your concrete crews don't cover them up.

Double-checking for accuracy

You should double-check all groundwork installations to make sure that nothing has been omitted. Adding plumbing to a system of underground plumbing before concrete is poured can be easy, but doing the same work after a floor is in becomes an onerous task.

A close inspection of all underground plumbing should be done before any concrete is poured. You should not schedule your concrete trucks and crews until you are comfortable with the fact that all needed

8-9
Protective cap placed on a toilet drain to prevent concrete from getting too close to the pipe

8-10 *Tub box*

plumbing has been properly installed. It is also worth saying that underground plumbing must be inspected and approved by your local code officer before concrete is poured. While I've never had it happen, I can imagine how upsetting it would be to pour a new concrete floor, only to find out that the plumbing inspection had not been done and the new floor had to be ripped up. Such a mistake would be expensive, time-consuming, and embarrassing. Don't send in your concrete workers until you have an official plumbing approval in hand.

Ensuring success

Your supervision and planning is what will allow you to complete the underground plumbing stage successfully. The basic plumbing involved in this stage of work is simple. Problems that occur are most often tied to negligence. If you take an active interest in inspecting groundworks before they are covered up, you shouldn't have to deal with many major problems.

9

Designing cost-effective DWV systems

One way to reduce the overall cost of plumbing work without sacrificing quality is to design cost-effective drain-waste-and-vent (DWV) systems. To do this, you need a good background knowledge of the local plumbing code. The DWV systems designed for most commercial buildings are created by architects and engineers. For single-family homes, these systems are almost always designed by a master plumber, but a knowledgeable general contractor can also draw up working plans for the plumbing system.

Most local jurisdictions require you to submit a set of plans and specifications for any type of extensive work before a plumbing permit will be issued. As a part of this submittal, a clear drawing of a riser diagram is normally required. Riser diagrams are line drawings that indicate the size and path of pipes planned for installation (Figs. 9-1 and 9-2). To the best of my knowledge, anyone can draw riser diagrams for use in residential applications; however, there may be some jurisdictions where a specific license or educational requirement is required for this job. Because the diagram has to be submitted to and approved by the code enforcement office before any work in the field is started, the general contractor can try creating the drawing. If the riser diagram is not in compliance with local codes, it will simply be rejected. Check with your local authorities before you assume it is okay to draw your own riser diagrams.

If you have no interest in actually drawing your riser diagrams, you might choose to sit down with your plumbing contractor and work out the design together. This can save you both time and money as your

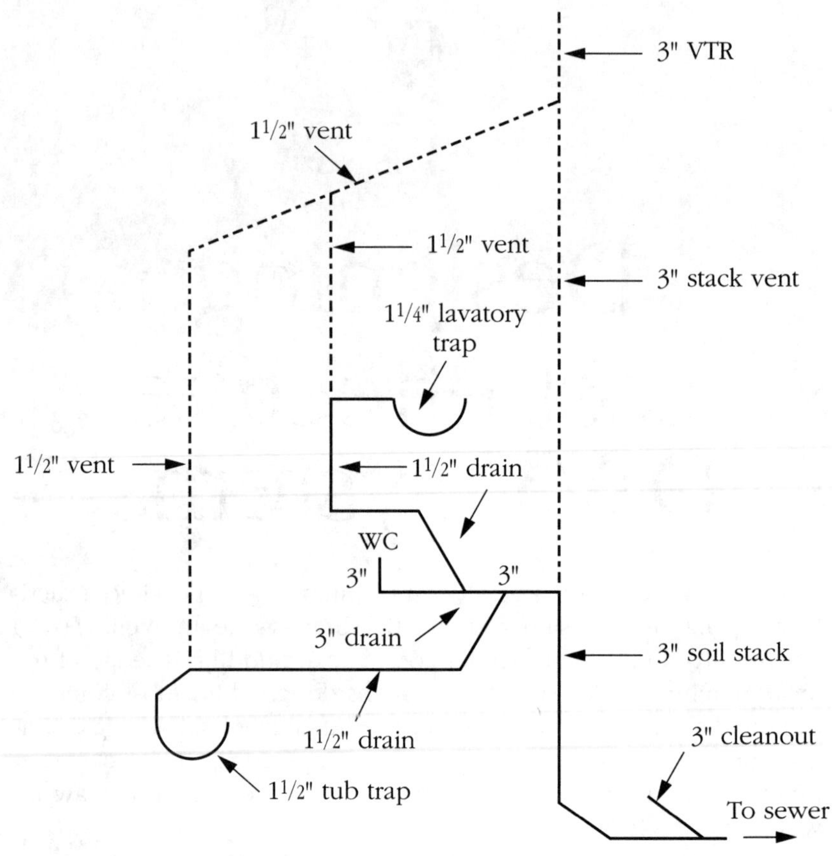

9-1 *DWV riser diagram*

project progresses. Like many general contractors, your time is valuable, so the savings of doing the drawings yourself may not be worth your effort unless you are dealing with a large job. The extent to which you wish to become involved with the design of your plumbing systems is up to you.

Some general contractors take an active interest in the design of plumbing systems for their jobs for reasons beyond saving a few bucks. These contractors want to see the proposed layout so that they can more efficiently plan their work. When you can see exactly what a plumber plans to do, you can look ahead to potential problems. Finding these problems while they are still on paper is a lot easier than facing them in the field. For example, a contractor might notice a problem with the installation of a steel flitch plate. This reinforcing steel might be blocking the only avenue open to a plumber's main drain. If this can be caught in the planning stage, there is a good

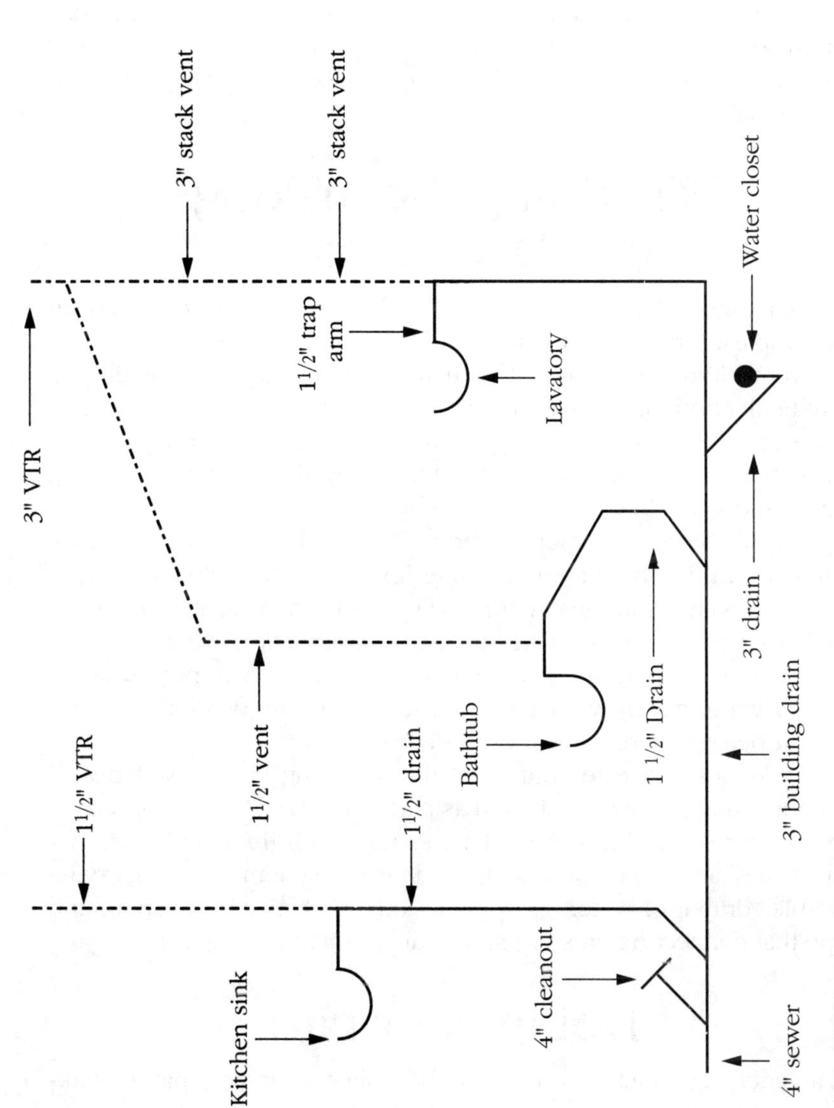

9-2 *DWV riser diagram*

chance you can find an acceptable alternative with minimal disruption to the job. Having such a plate installed when the plumber arrives to drill holes for pipes would present the contractor with a much more difficult set of circumstances.

The focus of this chapter is cost-effective DWV systems, but don't allow yourself to be blinded into looking only at the plumbing cost. The type of DWV system installed can affect other costs also, such as carpentry work. Always consider the entire job when designing your DWV system.

Describing the makings of a DWV system

The makings of a DWV system can be described in several levels. In the simplest terms, a DWV system is a group of pipes used to drain and vent plumbing fixtures. If you want to go into great detail, you could talk about individual fittings, hangers, and associated parts that make up such a system. This isn't necessary for the purpose of this book, but you do need to know a little more than just that a system drains and vents fixtures.

A DWV system, in more technical terms, is a series of pipes, located within the foundation and interior walls of a building, used to drain and vent plumbing fixtures (Fig. 9-3). The system's drain extends to a point up to 5 feet outside the foundation. Vents typically terminate into open air space above a roof line. Any pipe used to convey waste water, waste, sewage, sewer gas, air, or similar materials will be considered a part of the DWV system.

So as not to create confusion, the following are a few types of plumbing that are not considered as part of the DWV system: drainage systems for surface and ground water (although floor drains are because they are a part of a sanitary drainage system), pipes carrying potable (drinking) water, gas piping, and short pieces of tubing and pipe that connect fixtures to traps or drain arms (Fig. 9-4).

Doing your part

As a general contractor, you have to be willing to do your part in making cost-effective DWV systems possible. A plumber can't efficiently install plumbing if the building has not been designed or built with plumbing in mind. You may or may not be responsible for the design of your projects, but you are always responsible for the management

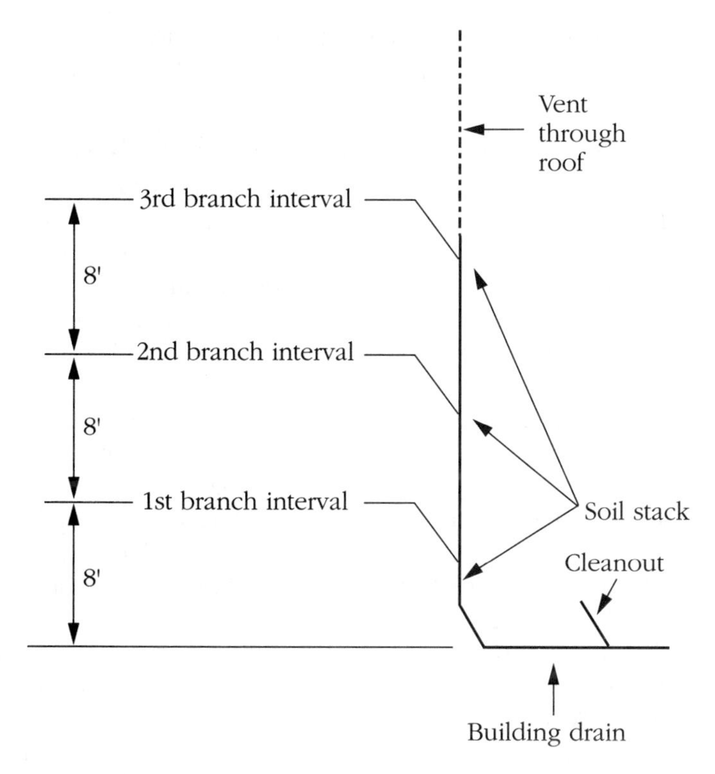

9-3 *Branch-interval detail*

9-4 *P-trap*

of a job. The following are some of the many factors that come into play when a plumber is trying to efficiently install plumbing.

Working on commercial buildings

If you are about to build a commercial building, you will probably have little to do with the overall design of the project. This work will most likely be done by engineers and architects. You might have some say in the design of a commercial remodeling job if the work is not too extensive; however, if it is, design professionals will probably have to work up the drawings. Typically, commercial contractors have little design responsibilities for their jobs.

When you are given a set of blueprints from which to work, it is your responsibility to carefully review them. It is not unusual for design professionals to design work on paper that is not practical in the field. These problems are usually minor and rarely abundant, but they may exist. Your careful scrutiny of the proposed plan can highlight potential problems before you face them in the field. This applies to all aspects of the design.

When you sit down with your plumbing contractor, you can go over the proposed plumbing design to see how it affects other phases of your work. Will there be a conflict with electrical conduits and underground plumbing pipes? Can the plumber work around the duct work? Are there any carpentry obstacles in the way of the plumbing? These are all questions that you and your plumber can work out in your office before a job has been started.

Dealing with residential construction

Residential construction varies from commercial construction not only with the types of materials used, but also with the methods used for installing them. There can also be a considerable difference in the details provided with working drawings for residential construction. For example, commercial blueprints detail a plumbing system right down to valve locations, while residential blueprints generally show only fixture locations. It is rare to get a set of residential plans that includes a plumbing diagram; therefore, a working plan for the plumbing installation must be created. The plumbing contractor is typically the one to create this plan.

The level of detail required for residential blueprints varies from location to location. Some areas require that these drawings be prepared by a licensed design professional, while a few areas don't require much more than a simple line drawing that shows the outline of a house. Because the level of detail and expertise required varies so greatly, consider your local conditions before taking an active role in the design of your own work.

I've worked in several states and have yet to encounter a situation where an architect's stamp was required on a set of residential blueprints, but I know there are places where certified plans are required for residential building and remodeling. If certification is required in your area, you will be limited to reviewing the work done by others in the design process, unless you are certified to produce the drawings.

Because my 20 years of experience along the East Coast has not shown a requirement for certified plans in residential work, I will base the following examples on my own experiences. I want to stress, however, that what I've done and am presently doing may not be appropriate in your local jurisdiction. It's always a good idea to check with your local authorities before taking too many steps in your own design process.

Assuming that you are creating a design for a new house, there is much that you can do to help your plumbing contractor. After your rough sketches have been transformed into a working plan by a drafting firm, you can call your plumber in to look over the drawings. This is when the two of you can plot a course for all of the plumbing. At this meeting, the plumber can point out any glaring problems with the plans, such as venting problems due to a vaulted ceiling or a sunken living room impairing the path of a major drain. Because everything is still in the planning stage, you can find viable options for any problems before your rough plan is turned into a blueprint. This type of preliminary planning makes everyone's job easier and can save you money.

Handling residential remodeling

Residential remodeling is considered to be a nightmare by some contractors, but I personally enjoy it. I've made very good money with remodeling jobs and enjoy the challenge because every job is different. However, the very things that I like about remodeling can create problems for contractors and plumbers. Unlike new construction, remodeling jobs are often more difficult because your options are limited. This is where your problem-solving skills can come in handy. The following are two typical examples of problems that occur in remodeling jobs.

Converting an attic

When you're called to a house to give an estimate on an attic conversion, you might have a pretty good idea of what you're getting into, but there are many hidden problems associated with these jobs. If the customer wants to add a bathroom in the converted attic space, how will the piping be run? Will the existing plumbing support an additional bathroom? You may be able to answer many of the plumbing questions yourself, but you would be wise to consult with your plumbing contractor. Once the plumber has inspected the job, a viable construction plan can be drawn up and plumbing installation plotted to allow for a trouble-free job.

Many remodeling contractors don't use formal blueprints in an effort to save some money. Although this may seem wise at the time, it can create problems that cost more than a good set of working plans. Once you have your plans drawn up, whether they are in your head, scribbled out on a napkin, or integrated into formal blueprints, you have to let your subcontractor know what you need and want.

There are several problems that come up frequently for plumbers who are working with attic conversions. One of the most common problem that arises in attic conversions is the lack of space in the floor joists, which makes it nearly impossible for a plumber to rough-in a drain for a toilet set at the normal floor level. The pipes cannot be concealed in the floor joists because of space constraints. This problem can be alleviated several other ways, including cutting out a joist and heading it off, raising the floor in the bathroom, or building a box in a closet below the toilet location. The latter is the best way to conceal piping for a problem hookup. Dealing with a problem of this type is much easier if you know of its existence prior to framing the job.

Toilet drains are not the only problems associated with attic conversions. Relocating existing vents in the attic can create plenty of confusion also. For example, if you have a 3-inch pipe rising through what will be the center of someone's new bedroom, you've got to find a way to move it without violating the local plumbing code. This can be quite a big problem, especially if turning a main vent to run horizontally violates code regulations. Your only option may be to go into the main part of the house to a point where the vent can be relocated to another part of the attic. The ripple effect of this type of trouble can be quite costly. That is why it should be worked out between you and your plumber well in advance.

Adding a bathroom in the basement

Basement bathrooms are popular remodeling projects in many parts of the country. If you are working on an effective plumbing and remodeling design for this type of work, you have many elements to consider. For example, where will you place the sump for a sewer pump, assuming that one is needed? Is there an existing drain under the floor that can be used to drain the bathroom? How will you get a 2-inch vent out to the open air? These are just a sample of the types of questions that may come up with such a project.

An experienced remodeling plumber can help you avoid most mistakes made by general contractors when planning the installation of a basement bath. Regardless of your level of experience, it never hurts to have a second set of trained eyes look over a job. Don't hesitate to ask for guidance from your plumber. Believe me, your plumber will probably be so thrilled to have a contractor who is actually concerned about creating an easy plumbing path that you will get outstanding service. Far too many contractors never consider the needs of their plumbers until it's too late.

Counting on your plumber

Your plumber's role in creating an effective DWV system will be a major part of your job's success or failure. Some plumbers can work wonders with difficult situations, while others often take an easy job and turn it into a horrible mess. You are on the right track if you have a competent plumbing contractor working with you.

Some plumbers don't like to be told what to do by general contractors, but it is your job to direct their actions. If you want to get involved in the layout of the plumbing systems for your jobs, it's best to wade in gently. Going to your plumber and expressing your intention to design your own plumbing layout is probably not a good idea. It would be far more effective to ask your plumber for help in laying out the system. If you explain that you want to be involved in the design so that you can help head off potential plumbing problems, your results should be much better.

There is no substitute for experience. Anyone with normal intelligence can learn the rules of the plumbing code. It is necessary to know and understand the code before you try to design a DWV system, though, and even that is not always enough. Experienced plumbers can predict problems before they happen. Don't get caught up in making a power play with your plumber on design issues, because you will probably lose. Let your plumber take the lead. It's okay to raise questions during the design process, but this is not the time to puff up and play boss. You are in the plumber's domain.

Once you and your plumbing contractor are comfortable with the idea of working together on design issues, a lot can be accomplished.

You can make design suggestions and your plumber can help point out potential problems with the design. By working as a team, the two of you can come up with a DWV system that is ideal for your job. Just remember, the plumber should be the boss during this phase.

During my many years in business, I've had a lot of general contractors tell me how to install plumbing. Sometimes their recommendations were very good, other times they were ridiculous. The point is, don't run off a good plumber by trying to be impressive with your newly learned knowledge of the plumbing code. Good jobs are made by working together as team players. You are the boss, but if you're a good one, you don't have to boss people around to get the desired results. A little mutual respect goes a long way.

Sizing and designing the DWV system

The technical side of sizing and designing a cost-effective DWV system is normally left to engineers, architects, and experienced plumbers. I don't doubt that there are general contractors who can do this work every bit as good as a plumber, but plumbers should be better equipped for the work. Bear in mind, however, not all plumbers are adept at this feat.

I've supervised thousands of plumbing installations over the years and have seen the work of hundreds of plumbers. One thing is for sure, it is difficult to find two plumbers who would plumb a job the same way. Every plumber has a personal style, as I call it, and it can get in the way. Many plumbers use more pipe and fittings to install a system than is required. They also waste a lot of time, which costs you money. To prevent this scenario, you have to put forth some extra effort.

Before you can tell if a plumber is being wasteful, you have to know what options are available. For example, in an effort to save money and cut the customer's price, a common vent (Fig. 9-5) can be installed to replace two individual vents (Fig. 9-6). Likewise, wet venting a water closet with a lavatory is a cost-effective alternative to installing an individual vent (Figs. 9-7 and 9-8). But how will you know if this is acceptable and when it should be done?

In order to understand all the rules pertaining to the plumbing trade, you have to learn and understand the plumbing code. I warn you, the code can be cryptic. Professional plumbers often don't understand the very code with which they work, so don't think that

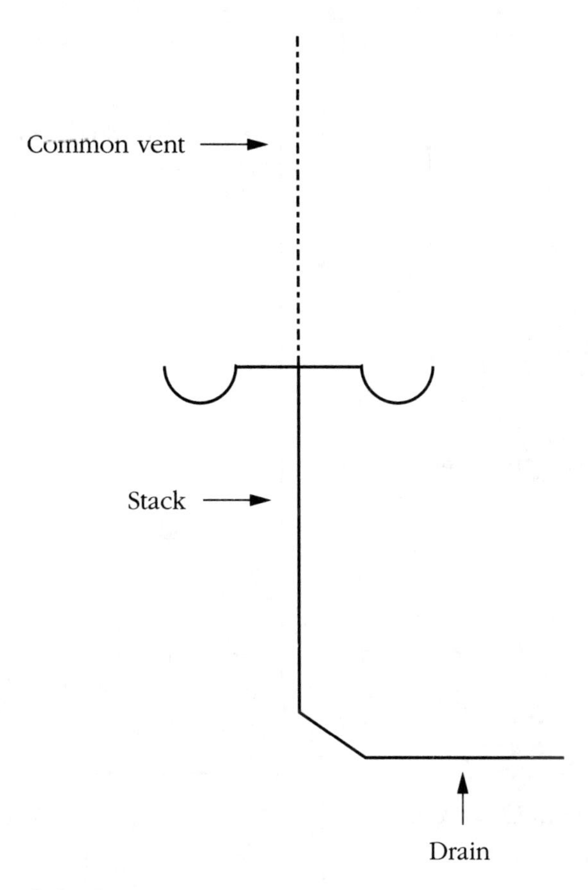

9-5 *Common vent*

you're going to sit down for some light reading and walk away with a wealth of knowledge pertaining to the plumbing code.

A person who has a solid understanding of the plumbing code can find a multitude of ways to reduce the time and expense of installing a DWV system, but the code is not the only consideration. Common sense can play a big part in the creation of a good system. You must know what the code will allow you to do; then it is up to you to determine the best design.

Falling into the designing rut

A lot of plumbers fall into a designing rut when it comes to DWV systems. They learn one way of designing the system and rarely stray from it. I've seen this time and time again. When asked why they

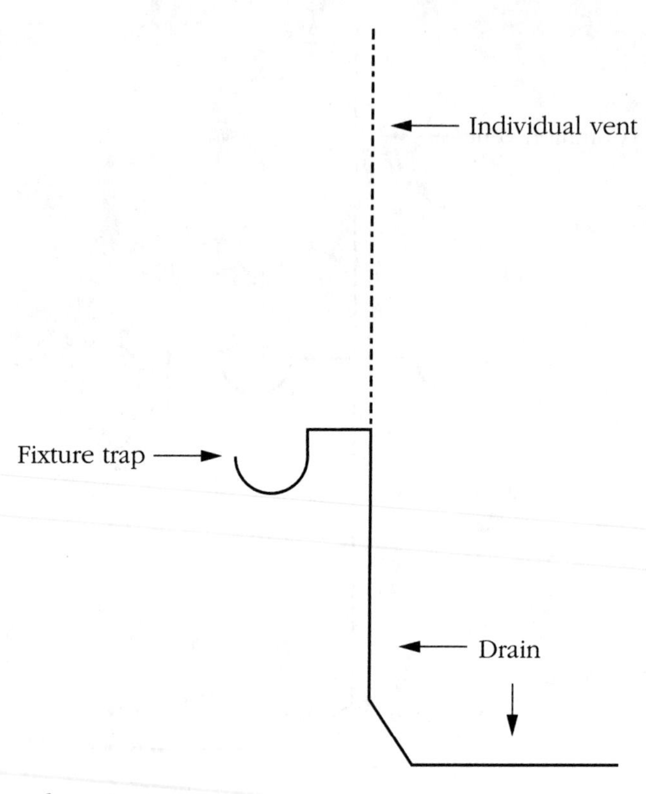

9-6 *Individual vent*

don't do the job like this or that, their reply is, "We've always done it this way." Doing it the same way all the time is not always good enough. If your plumber is unwilling to branch out into new styles and designs, you might be better off with a new plumber.

I've seen plumbers run individual vents for every fixture in a house. By code, this is an acceptable practice, but from a business point of view, it is a waste of money. Most codes do require that all fixtures be vented, but this doesn't mean that an individual vent has to extend from each one. The use of common vents, battery vents, wet vents, and a multitude of other types of vents can reduce the amount of time and pipe needed to do a job (Figs. 9-9 through 9-18). This can translate into monetary savings.

Most of the plumbers who are in designing ruts have very little knowledge of the plumbing code. They know the basics, but that's where it ends. For a plumber to design and install a sleek DWV system that is cost-effective, full use of the code must be employed. This pertains to the sizing of pipe, the choosing of appropriate fittings, and

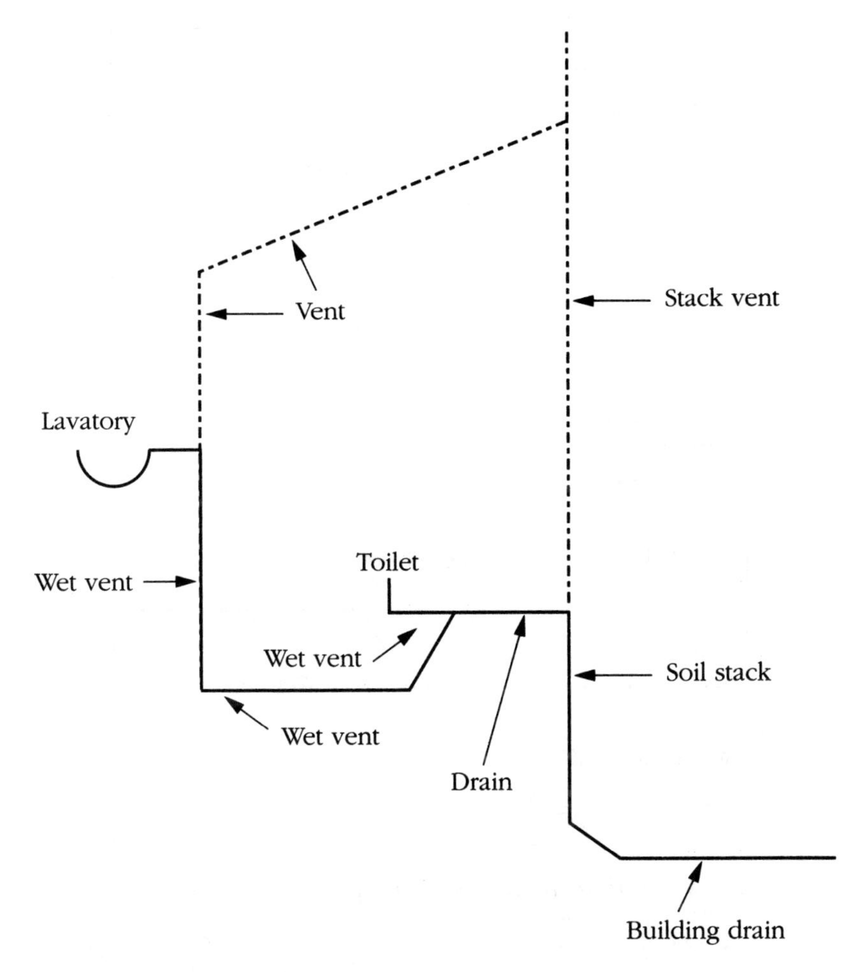

9-7 *Wet venting a toilet with a lavatory*

the employment of the most effective methods for actual installation. Any plumber who is not fluent in code requirements is handicapped when it comes to designing and building the ultimate DWV system.

Sizing the pipe

Pipe sizing for a DWV system is really quite easy. The plumbing code assigns numbers, represented as fixture units, to specific types of fixtures. For example, the minimum size of a drain pipe is determined by its length, the amount of its pitch, and the total number of fixture units being served by it. Anyone who can use tables and charts can size this type of piping. The plumbing code provides all the data

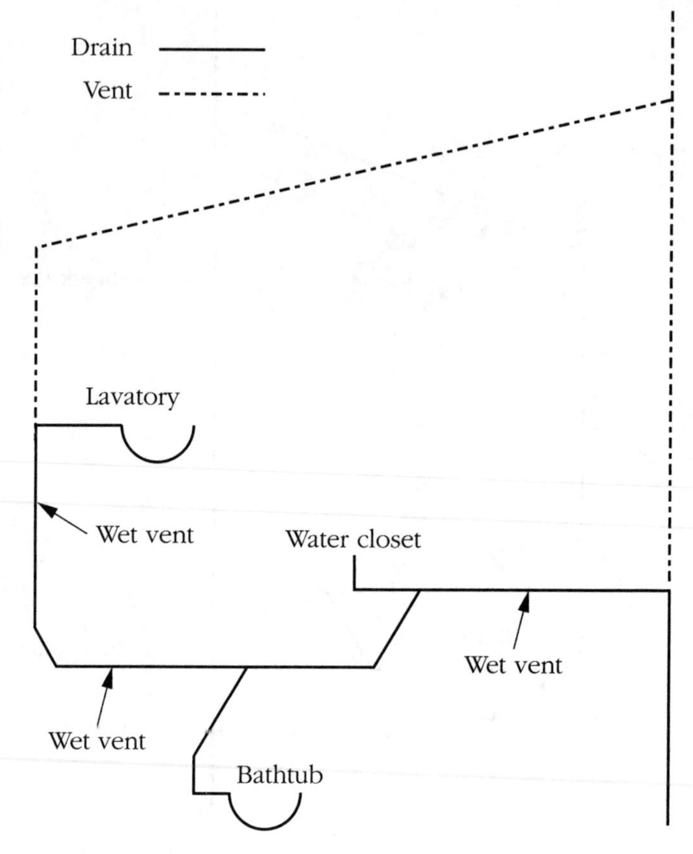

9-8 *Wet venting a bathroom group*

needed to easily determine a minimum pipe size. Anyone who is willing to spend a hour or two doing sample sizing exercises can learn how to size a pipe for a DWV system.

As simple as it is to size pipe, many plumbers try to avoid sizing exercises. They would rather use a 2-inch pipe to be safe than to find out that a 1½-inch pipe will do just fine. The cost difference in one pipe size is not a lot when you're just looking at a 1-foot section, but it adds up as hundreds of feet of pipe are installed. In a large job, such oversizing could cost a lot of money.

There are times when it is worth a little extra money to run a pipe that is larger than the minimum size required by code. One example, in my opinion, is the drain for a kitchen sink. I like to run that drain with a 2-inch pipe even though a smaller one is allowed because food particles and grease are continuously dumped down it.

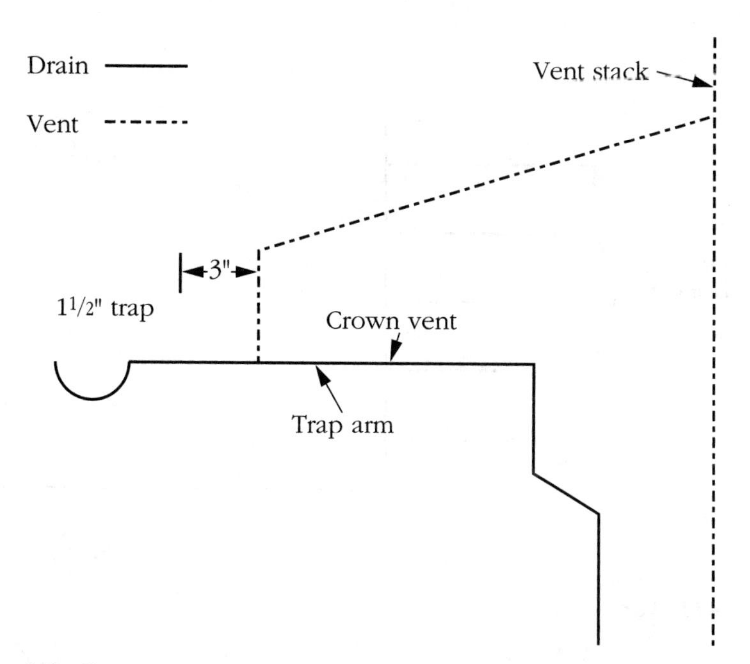

9-9 *Crown venting*

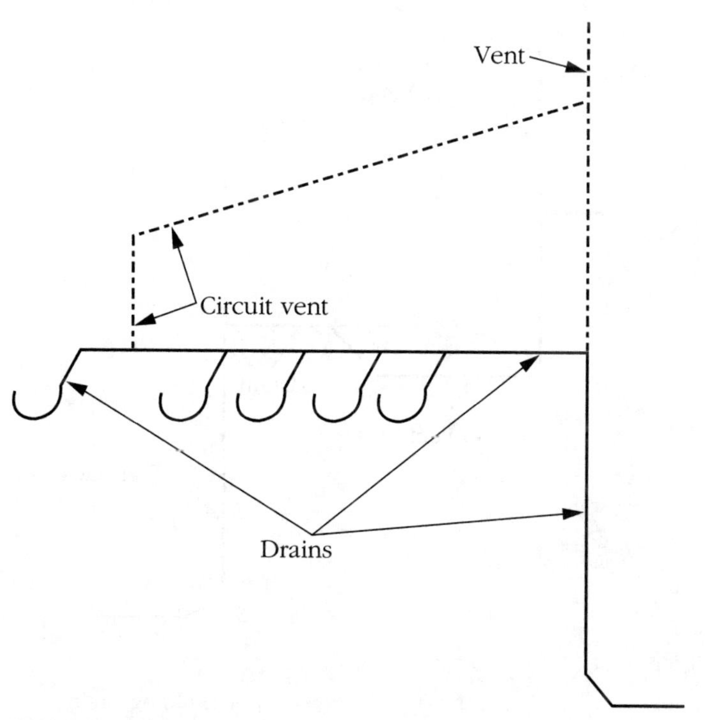

9-10 *Circuit vent*

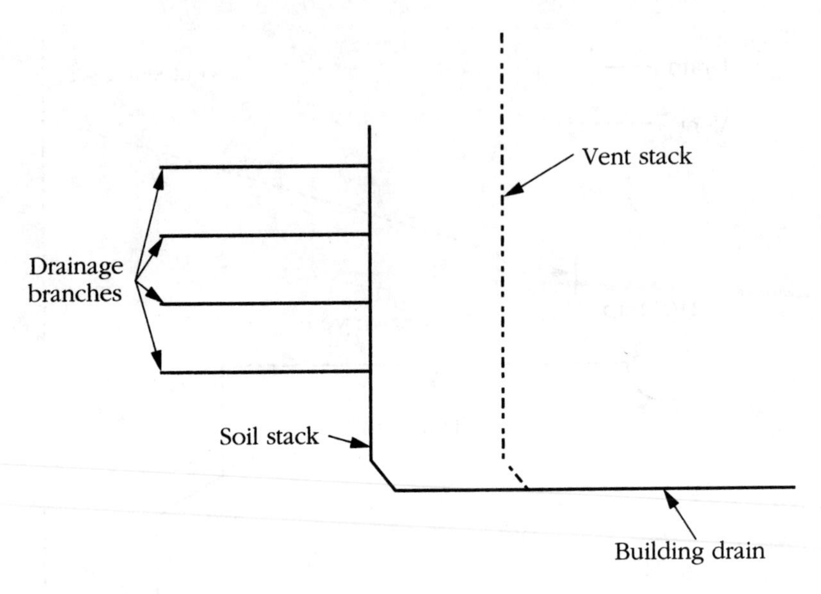

9-11 *Vent stack*

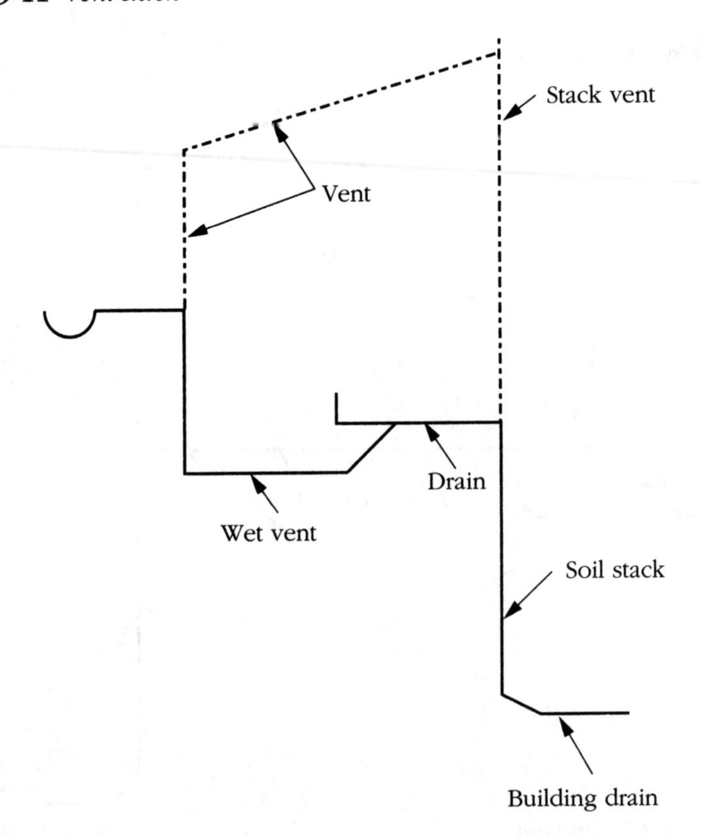

9-12 *Stack vent*

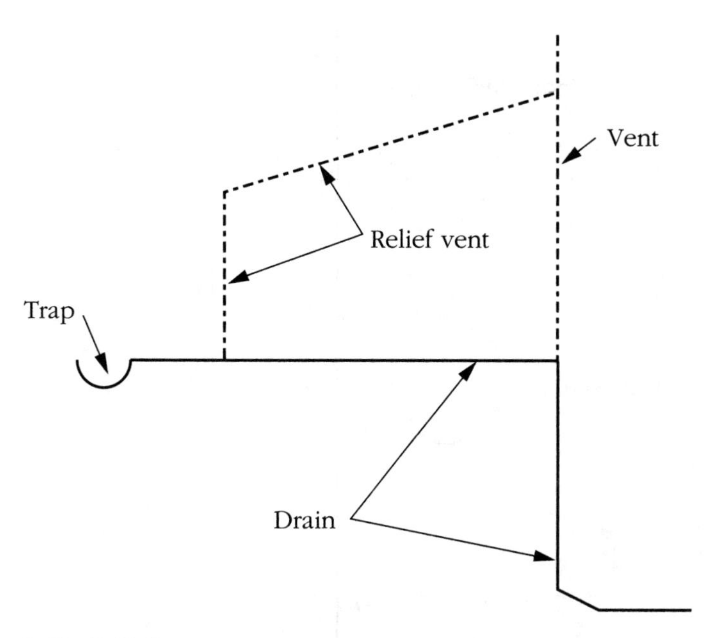

9-13 *Relief vent*

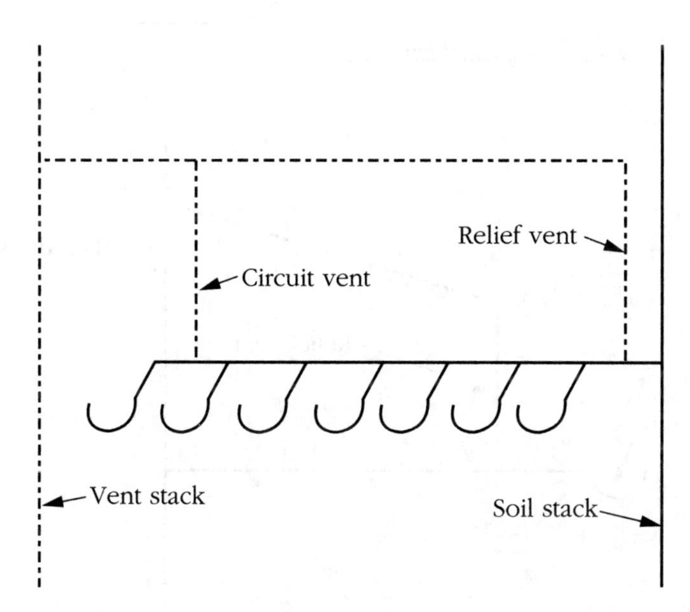

9-14 *Circuit vent with a relief vent*

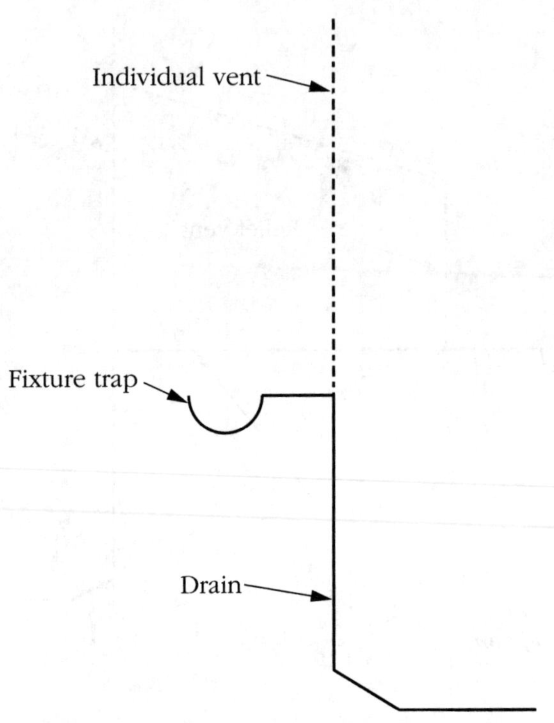

9-15 *Individual vent riser diagram*

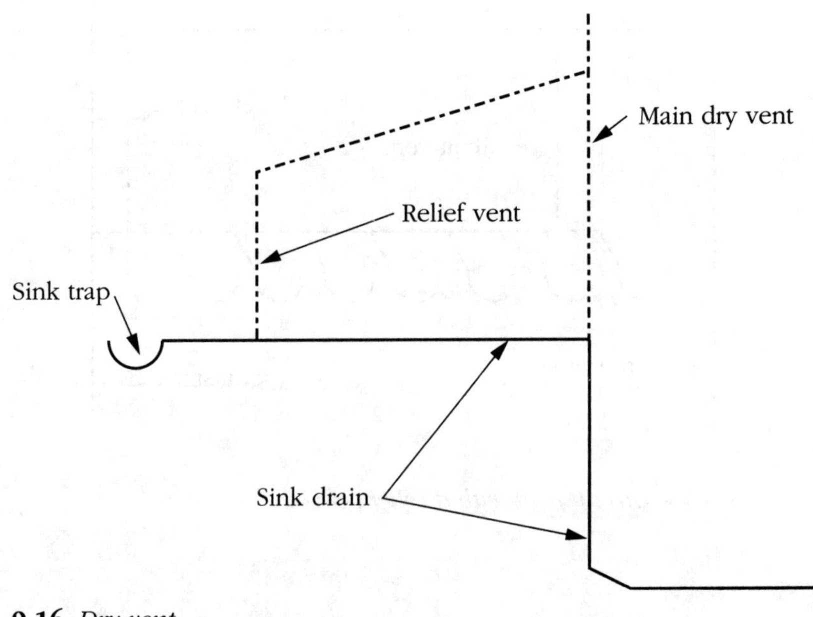

9-16 *Dry vent*

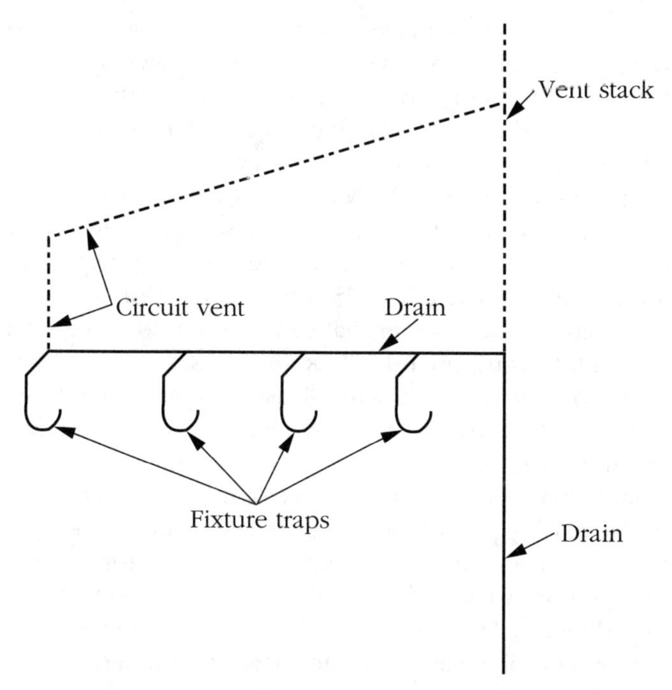

9-17 *Circuit venting multiple traps*

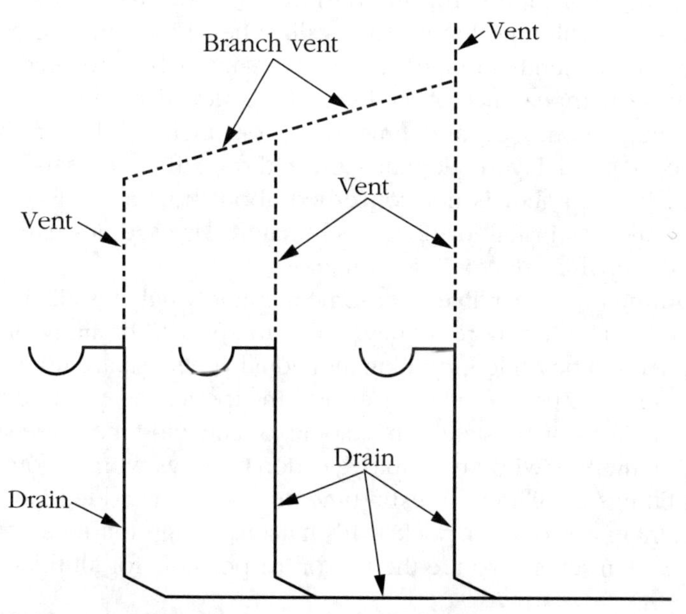

9-18 *Branch vent*

Sewers are another place where I like to oversize pipes. The plumbing code that I work with most often allows up to two toilets to discharge into a 3-inch pipe. Depending upon the number of toilets being installed and the total fixture-unit count, it is possible to run a 3-inch sewer for some houses. I try never to do this, however, because if the homeowner ever wants to add a third toilet, the sewer has to be replaced with a larger one or a separate sewer must be installed. On an average residential job, the cost of upgrading the sewer to a 4-inch pipe is less than $75. For this kind of money, I would rather give customers a sewer that can accommodate new additions.

I've just told you how I don't like 3-inch sewers, but now I'm going to tell you about a time when I used them for a 365-unit townhouse project. If I conservatively say that the cost difference between a 3-inch and 4-inch sewer averages $50, how much money could be saved on a 365-unit townhouse development by using the smaller sewer? The savings would be $18,250. This is not pocket change. If you were to bid such a project with the cost of a 4-inch sewer when your competitors were bidding with 3-inch sewers, would you win the job? I doubt it. There are times, like in this example, when going with a minimum pipe size is of paramount importance.

Reviewing the system's general layout

The general layout of a DWV system can affect its cost a great deal. For example, if pipe is run in a random method, more pipe will be used, and it will take longer to install. When pipe runs are consolidated and efficiently installed, less material and labor are needed, and the physical appearance of the job is more attractive. This means that you can save money and time while producing a better job. Of course, a good layout depends on a thoughtful mind and skilled hands. If a plumber is not concerned about how a job looks, how long it takes to install, or how much material is used, it's hard to say what the final product will look like.

During my career, I've seen some incredibly ugly installations that cost someone a lot more money. This can apply to drilling holes that don't need to be drilled, running individual vents that are not needed, and a lot of other installation errors. Inexperienced plumbers typically make these mistakes, but some of the old-timers are just as guilty. Plumbers who are employees don't always worry about using extra fittings or billing out extra time on a job. The added job cost is not coming out of their pocket. It's hard to design the most efficient DWV system unless you're the person responsible for all the costs.

As a general contractor, you may not have the experience to size and design your own DWV systems, but this can change. If you are willing to spend some time studying the plumbing code, you can learn how to size pipes and design a cost-efficient plumbing system. Once you have working knowledge of the code, you can have more meaningful conversations with your plumbing contractor.

Decoding the plumbing code

Learning how to use the plumbing code to your advantage will require time and effort on your part. The code is not always easy to understand. If you feel uncomfortable working your way through a code book on your own, look into taking a code class. I used to teach a code class at a local college that met two evenings a week for about six weeks. The cost for this class was around $100. Most of the students were plumbers who were preparing to sit for their licensing exams.

During my classes, I taught all aspects of the plumbing code. I involved the class in sizing exercises and similar projects, so the students were able to learn by doing. I feel this is the best way to gain a full understanding of the code. There is nothing to say that you can't teach yourself the code, but a good code class can help you reach your goal quicker.

I said earlier that experience was needed to effectively size and design DWV systems. As a general contractor, you should already possess some plumbing experience. You may have never installed a pipe, but you've probably seen a lot of plumbing installations, which counts as experience. This isn't the same experience as a plumber's, but it is more than the norm.

Once you learn how to use the plumbing code, you can start experimenting with what you know. It is wise to run your ideas by a good plumber before you put them into practice. When your plumber approves your ideas, you know you're on the right track. The more you do this, the better you will get at it and before you know it, you might just be laying out your own plumbing designs.

10

Installing cost-effective water-distribution systems

Cost-effective water-distribution systems are just as important to a money-making job as an efficient DWV system. Fortunately, laying out an effective water-distribution system is easier than designing a DWV system. If you are working with residential plumbing, installing a legal, cost-effective water-distribution system will not be very difficult.

If your plumbers are still using copper tubing for the potable water systems, you could be spending more money than is needed. You can cut your copper costs considerably when you put a little extra time into planning a good system by using proper pipe sizing and layout. Because copper tubing is expensive, you could make your jobs more profitable by using smaller sizes and less of it.

I've found that most plumbers are more comfortable with the design and installation of water systems than they are with DWV systems. This isn't difficult to understand because the process with water systems is easier. Although the pipe sizing does not seem to be a problem for most plumbers, they are still wasteful in their layouts, costing the general contractor extra money with their roundabout pipe routes.

If you've been a contractor for long, you've seen many plumbing rough-ins. Whether you've paid attention to them or not, I'm going to give you some examples of both the right and wrong ways to install water pipes. By wrong ways, I'm not referring to installations that violate the plumbing code. I'm calling bad installations those that cost extra money. Obviously, good installations save you money.

Sizing the pipe

When you are sizing potable water pipes for residential work, there are some very simple rule-of-thumb figures to use. These numbers won't always prove true, but they will often enough. Keep in mind that not all plumbing codes are alike and that the rules for your local jurisdiction may conflict with my examples.

As a rule, not more than two plumbing fixtures may be served by ½-inch pipe. Typically, a ¾-inch pipe is large enough to use as a water service and main water-distribution pipe. Water heaters should have ¾-inch pipe installed for both the supply and discharge pipes. Knowing just what I've told you in this paragraph, you could size a legal water-distribution system.

To use the information I've just given you, you would run ¾-inch pipe as your main water lines, both hot and cold. As you neared the end of your run, you would downsize to ½-inch pipe for the last two fixtures on each run. That's really all there is to it. Of course, this rule-of-thumb sizing is not always appropriate. Some fixtures require only a ⅜-inch water supply. Most plumbers run ½-inch pipe to these fixtures, but some do install smaller supplies. This practice has become more common as the use of manifold systems has grown.

If you want to size water pipes with maximum efficiency, you will have to consult your local code book. There you will find tables, charts, or text that detail minimum pipe sizes (Table 10-1). The sizing is based on such factors as the length of a pipe run, the working water pressure rating, fixture-unit ratings (Table 10-2), and minimum supply sizes. It might sound a little complicated, but it really isn't. Once you learn to properly use the information in the code book, sizing becomes a simple process.

Comparing material costs

Material choices can have a significant effect on the cost of your next plumbing job. For instance, by comparing the cost of copper tubing with that of polybutylene (PB), you can see very quickly that copper is a lot more expensive. Even though the fittings used with PB tubing may cost more than the ones used with copper tubing, the overall cost of materials for a job with PB pipe are normally less.

Table 10-1. Recommended minimum sizes for fixture supply pipes

Fixture	Minimum pipe size (in inches)
Bathtub	½
Bidet	⅜
Dishwasher	½
Hose bibb	½
Kitchen sink	½
Laundry tub	½
Lavatory	⅜
Shower	½
Water closet (two-piece)	⅜
Water closet (one-piece)	½

Table 10-2. Common fixture-unit for water distribution

Fixture	Hot	Cold	Total
Bathtub	3	6	8
Bidet	1.5	1.5	2
Kitchen sink	1.5	1.5	2
Laundry tub	2	2	3
Lavatory	1.5	1.5	2
Shower	3	3	4
Water closet (two-piece)	0	5	5

There are many types of approved water-distribution pipes on the market. CPVC plastic piping is sometimes used by homeowners, but rarely by professional plumbers. Of all the types of approved pipes for use, copper and PB tubing see the most use in professional installations of interior water-distribution systems.

I've used copper tubing since the beginning of my plumbing career 20 years ago. I used PB pipe when it was first available in my area, but not for long because of the initial problems associated with its fittings. As technology improved the fittings for PB pipe, I returned to it. Today, I use much more PB pipe than copper. You may have noticed that I switch between the word tubing and pipe. Technically, the material used for most residential work is tubing, but almost all plumbers call it pipe. Forgive me for using the wrong term, but old habits die hard.

Some people still resist the use of PB pipe, mostly in part due to bad publicity. I think some of the reason for a slow acceptance of PB pipe is that old-school plumbers are not fast to change from their old materials. The idea of using a crimping tool instead of a torch to make pipe joints is just more than some plumbers can stand.

Based on my professional experience and the documentation I've studied, I feel confident in recommending the use of PB pipe. With this material, it is necessary for the plumber to make the joints properly; if not, they can blow apart. The same problem is possible if the plumber is not experienced in working with PB pipe. As long as you have a qualified plumber, I think you will be very happy with PB pipe. If you select this pipe, you should save money on both labor and material costs.

Laying out the potable water system

The layout of a potable water system has a lot to do with its cost (Fig. 10-1). This is where most plumbers make costly mistakes. I've found that plumbers generally size water pipes correctly, but often route the piping in a manner that is not cost-effective. It is my guess that plumbers make this type of mistake out of a lack of interest in saving money. Because most plumbers are employees, they are not paying the bills for their time or materials. This leads to a setting where excessive costs are common. The following are examples of inappropriate potable water system layouts.

Installing the piping for an upstairs bath

Suppose you are contracted to install an upstairs bathroom with the piping coming up from below and full access to install the pipe any way you choose. How are you going to cost-effectively install them? Many plumbers will pipe this bathroom with vertical risers for each fixture. This may occasionally be the most effective route to take, but it normally isn't. It is often better to bring two risers into a partition wall and feed the pipes through the wall. This typically saves time for the plumber as well as saves money on the pipe.

When risers are brought into the upstairs partition wall, the plumber is not working from a ladder for nearly as long, which improves production. Furthermore, by consolidating the pipes in a common wall, it is often possible to use less pipe. This all adds up to less expense.

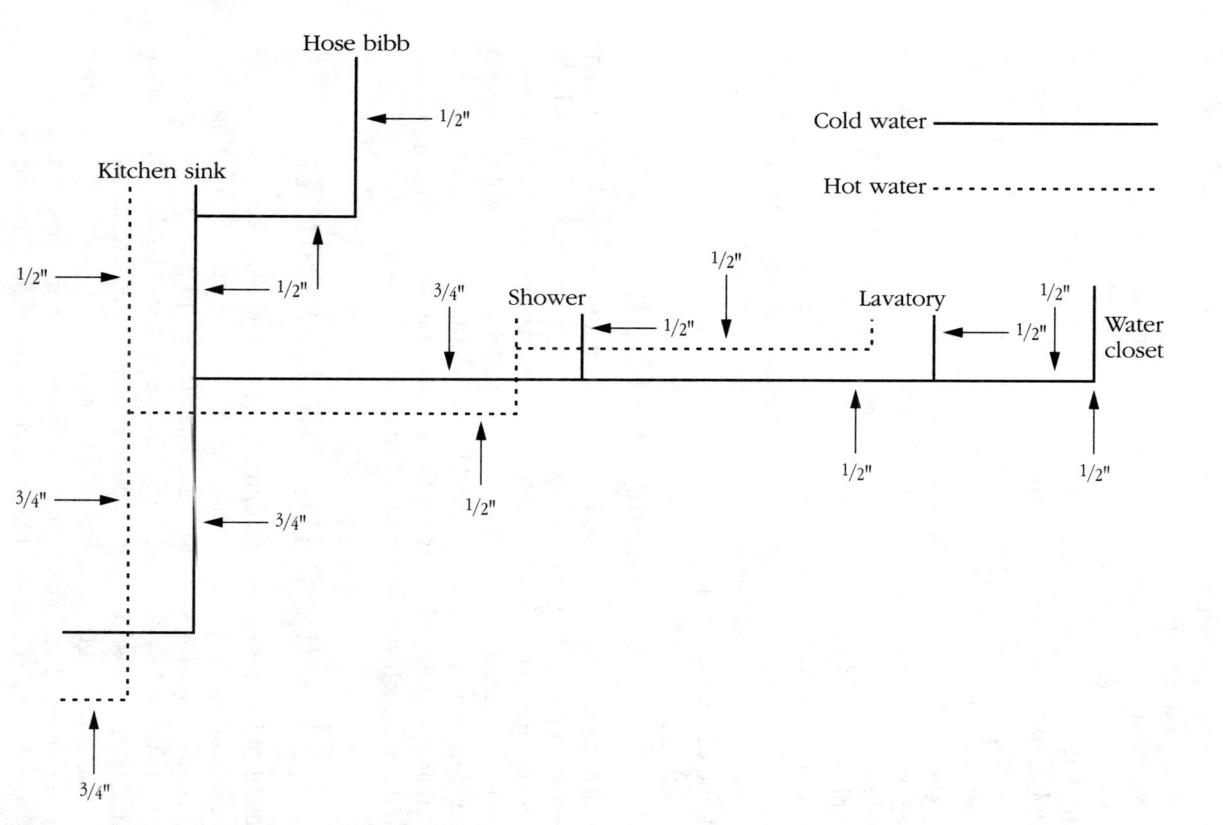

10-1 *An efficient water supply diagram*

Some plumbers make a mistake when sizing pipe for a single bathroom. A typical bathroom contains a bathing unit, a toilet, and a lavatory. This creates a need for two hot-water supplies and three cold-water lines. What does this tell us? Think back to the rule-of-thumb sizing examples I gave you earlier.

According to the aforementioned information, two plumbing fixtures can be served by a ½-inch pipe. This means that a ½-inch pipe is all that is needed to supply a single bathroom with hot water. The supply pipe will have to originate from a ¾-inch pipe, but running ½-inch pipe from a basement to an upstairs bath will save more money than if a ¾-inch riser was used.

The cold-water demand for the full bath requires a ¾-inch riser. Once the riser reached the first cold-water fixture, it can tee off with a ½-inch supply for the other two water supplies. I've seen a lot of plumbers run ¾-inch risers for both the hot and cold water. This is good because the customer is getting a higher volume of water, but it's also bad because the plumber is spending more for the pipe. Only the customer can determine which is more important, giving more volume or saving money.

Adding a half-bath

A half-bath requires only one hot-water pipe and two cold. This means that ½-inch pipe is the only supply size required for the half-bath. If you want to carry pipe sizing to extremes, you can get by with a smaller pipe for individual fixture supplies. A ½-inch riser would still be needed, but a ⅜-inch supply could be used for the second fixture. If you're using a manifold system, where every fixture has an independent supply, you could use ⅜-inch supply for all the fixtures in the half-bath.

Using a manifold system

A manifold system is to plumbing what a circuit-breaker box is to the electrical system. Basically, a manifold is one central location from which all plumbing supplies originate. There are pros and cons to using a manifold system. Although these systems have only become popular in the last few years, they appear to be here to stay.

The advantages to a manifold system apply to both consumers and plumbers. For a consumer, the biggest advantage is that it allows all supplies to plumbing fixtures to be controlled individually. If a person wants to shut down one water line, it can be done. When three water lines need to be shut down, they can be. This is not totally different from the options possible with shutoffs placed near each fixture, but it can be more convenient. Because many plumbers don't install cut-off valves for bathing units, a manifold system makes cutting the water off to these fixtures simple, fast, and easy.

Aside from flexibility in controlling the flow of water to individual fixtures, there are not many other advantages for homeowners when discussing manifold systems. Every individual may have some personal preference for a manifold. I plumbed my house with a manifold system because I like being able to control all of my plumbing supplies individually.

From a plumber's perspective, the advantages of a manifold system are a little different. Manifolds allow plumbers to run smaller pipes to individual fixtures. The savings in cost from using smaller pipe is offset by the expense of the manifold; therefore, the savings are not worth considering.

Running all water supplies from a central source makes it easy to size water pipes. That also means that the minimum fixture supply can be used to establish all supply-pipe sizes. This is an advantage, but it is a small one.

Probably the best advantage for the plumbing contractor is the time saved when installing a manifold system. This is especially true when PB pipe is used. In fact, manifolds are rarely practical unless PB piping is used. With this flexible pipe, plumbing the water lines in a house can be as fast as pulling electrical wires.

The only disadvantage I can think of to using a manifold system is cost. The expense of a manifold is one that is not encountered with a traditional water-distribution system. However, the time saved in labor can offset this cost.

Learning the basics

If you learn the basics of water-distribution systems, it will not take you long to learn how to cut down on your plumbing costs (Figs. 10-2 and 10-3). When you gain enough knowledge, you will be able to quickly spot a good system. Being able to separate the good systems from the bad is the first step in becoming skilled in the design and implementation of a cost-effective system.

You should learn as much as you can from the plumbing code about water pipes and their installation. You should find code issues pertaining to water pipes much easier to understand. With a little practice, you can develop strong skills in creating efficient water systems.

As with most things, practice is the key to success. As you begin to pay attention to water systems, you will be able to pick out the

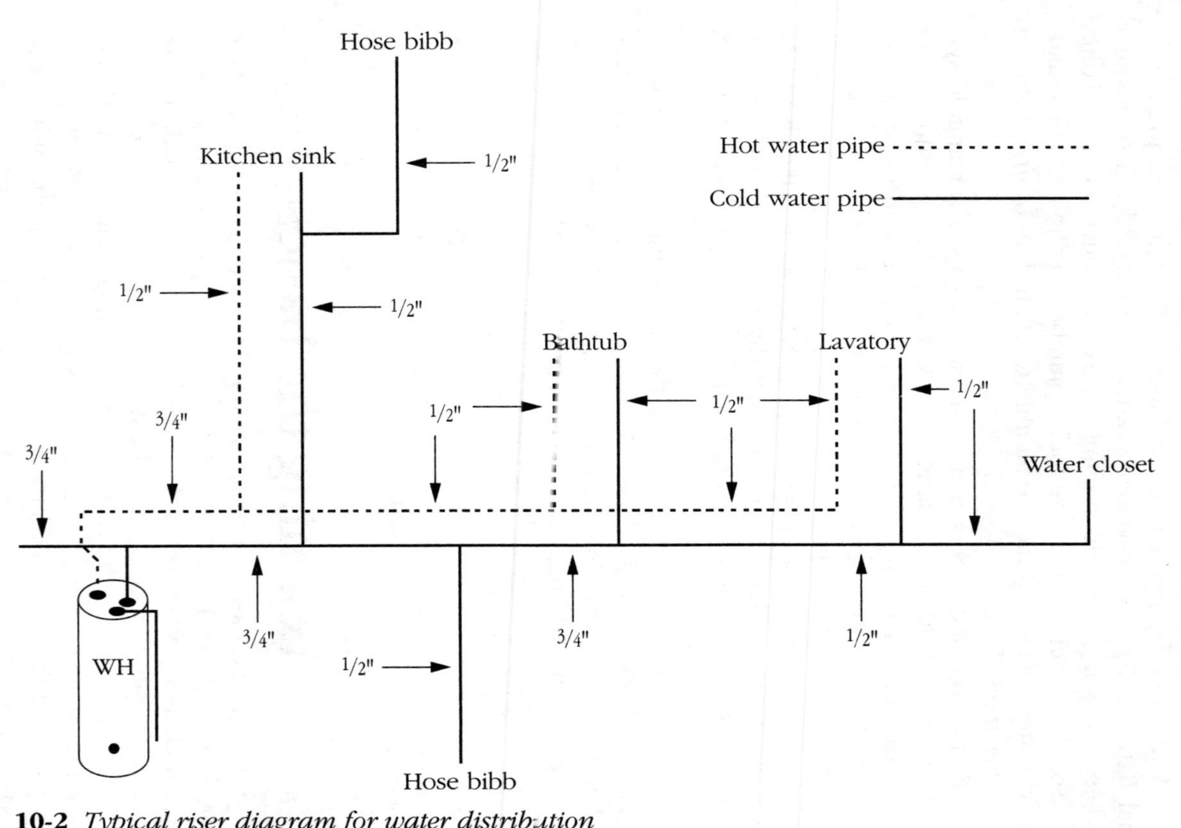

10-2 *Typical riser diagram for water distribution*

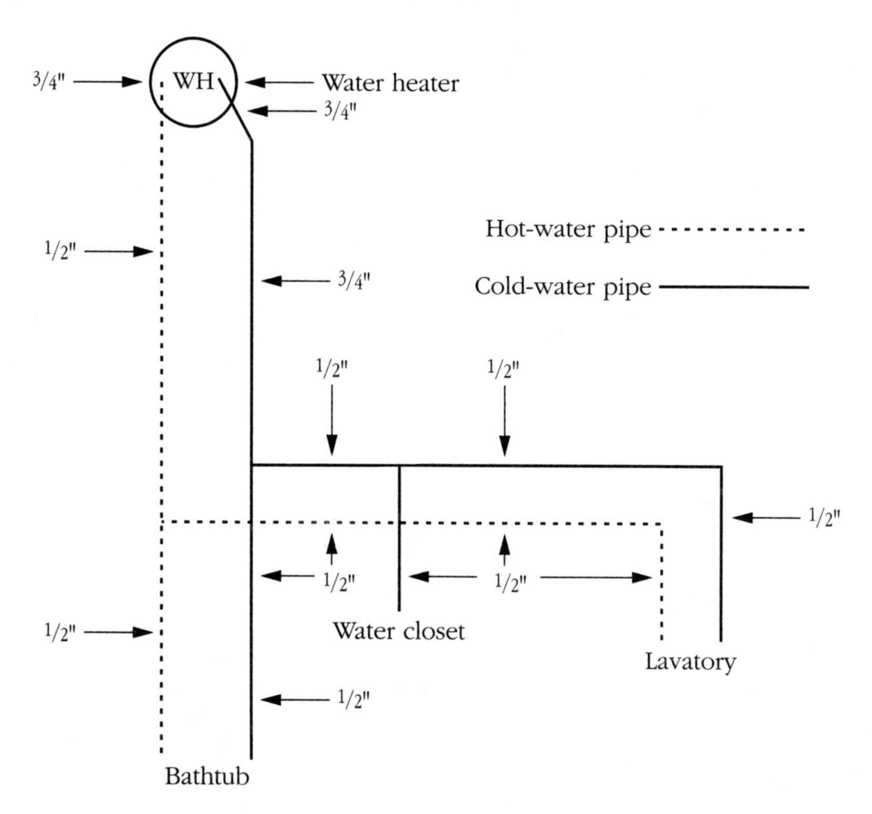

10-3 *Typical riser diagram for water distribution*

flaws. The more you look, the more you'll see. When you get an eye for detail, you are ready to start your own designs. You may never design a complete system, but your knowledge of the code and good installation practices will make it possible for you to work better with your plumbing contractor.

11

Tapping into water sources and services

Water sources and services are needed for all plumbing systems, but they are often taken for granted. People tend to assume that water is abundant and always available. This just isn't true. As a general contractor, you might come across a job in which water becomes a very big problem. Can you imagine building a house and then finding that no water source is available for it? Although I've never heard of such a problem, it could happen.

If your work is done outside urban environments, you are probably accustomed to working with water wells and pumps, which are quite common in rural areas. Contractors who work exclusively within city limits don't have much occasion to deal with such units. Whether you are working with a private or municipal water source, you will have to work with water services. These are the pipes that convey water from a water source to a water-distribution system within a building.

Water is an essential element in a plumbing system. If there is no water available, you've got a serious problem. The potential problems associated with water sources could scare you into early aging. Fortunately, there is usually some reasonable way to get water to projects without unreasonable costs. Bear in mind, though, if you get caught, even just once, without a suitable water source, you could be out of business. Because this is such an important issue, this entire chapter is devoted to discussing the various types of water sources and services.

Accessing city water

A great number of plumbing systems are served by city water. This dependable source of potable water is easy for plumbers to access. When there is a municipal water main to tap into, there is no doubt that an abundant quantity of potable water will be available.

Contractors who have access to city water are in a more secure position than contractors who rely on wells. When wells are drilled, there is no guarantee that water will be found. Even if a vein of water is hit, there is no assurance that it will produce a suitable quantity of water for an ongoing period of time. Contractors with city water access are lucky they don't have to worry about a well, but there is a price to pay for this security.

The price is taken out monthly in the form of water bills associated with municipal water supplies. Such bills are nonexistent for people who get their water from private wells. Drilling a well is not an inexpensive job, but once it is done, there is no ongoing payment required for the water. When you consider the monthly cost of city water on a long-term basis, a well doesn't look like a bad idea. However, many cities and towns require water and sewer hookups. If the connections are available, properties are required to hook up and pay the monthly bill. This may not be true in all areas, but it has been the case in areas where I have lived and worked.

People normally have wells because no municipal supply is available. Not many people I know would opt for a well over a municipal water supply. One reason to avoid the municipal water supply could be the cost, but this might be deceiving. The fee charged to tap into a city water main can be nearly as much as the cost to drill a well, but keep in mind that this fee is a one-time expense. Once you tap into the municipal water supply, monthly bills follow for water usage. Drilling the well is another one-time expense, but, again, there is no ongoing expense with the exception of electricity to run the pump. So which is really less expensive? When a stiff tap fee is charged for a city hookup, a well is more economical, but not always as dependable. Be sure to check your jurisdiction requirements with regard to mandated city hookups, when they are available.

It's a common fact that most people choose to use city water when it is available. If your work area is served by water mains, they are your primary water source. A well might be dealt with occasionally, but the water mains will provide water to most of your jobs.

Tapping into municipal water connections

Municipal water connections can be quite simple, but they can also be complicated and expensive. The factor that determines the connection's complexity comes from the municipality. If you have to tie in directly to a large water main that happens to be buried in the middle of a paved street, the connection will be very difficult to make. When the connection is made on the house side of a water meter, the connection is simple. As the general contractor, you had better determine this type of information before quoting a price to a customer. The difference in cost for a curbside hookup and a center-street hookup is likely to be astronomical.

Making curbside hookups

Curb hookups are easy to make, but can still be expensive. Most cities charge a connection fee for the privilege of tapping into a municipal water service. This fee can reach into thousands of dollars. If a contractor has not allotted the money for this fee in the cost of a job, the profit picture can turn gloomy.

After moving to Maine, I have not made many curbside connections. Most of the jobs I have worked on in Maine have gotten their water from private wells. The last curbside connections I made were in Virginia. This was about eight years ago, and the cost for connection at that time was about $2300. I imagine the fee has gone up, and I should add that not all cities charge similar fees. Some city's fees may be very reasonable; others are extremely high. Keep in mind that the connection fee is only for the privilege of making a connection—it does not include the labor or material required to do the actual connection. The price for this work is usually set by the plumbers, but sometimes the site contractor will give the quote.

One good thing about curbside hookups is that the connection point is right at the property line for the building being served. To make a connection with this type of arrangement, a trench has to be dug, and pipe has to be run from the connection point to the building. The excavation work is the most expensive part of the installation. A plumber's labor for running a water service pipe shouldn't be much at all. When the total cost of a connection fee, excavation work, and labor is added together, you could be looking at several thousand dollars. Even at this, your cost is much less than if you have to make a street connection.

Making street connections

Street connections can cost so much that the figure will scare you. If your locality requires you to take responsibility for all of the work involved with a street connection, proceed with extreme caution. The tap fee for a street connection may be the same amount as would be charged for a curbside hookup, but the cost of installation will be a great deal more.

I've only made one street connection in the past few years, but that one was enough! The rules and regulations pertaining to street connections vary. In my experience, the contractor has been required to accept responsibility for all work involved. This may not be the case in all areas, but it has proved to be true on my jobs. If you take full responsibility for a street connection, you had better be darn sure of what you are getting into. Let me use my last street connection as an example.

When I was required to provide a street connection, my first order of business was to pay the connection fee. In the little town where I was working, this fee was less than $2000, but that fee was only the beginning.

After the tap fee was paid, I had to arrange for subcontractors to cut the paved road surface and dig up the road bed so that my plumbers could make the tap on the water main. Once we had gotten the tap made and the pipe extended to my customer's property line, the road had to be repaired. This is where the job got very sticky.

In my particular case, the local authorities were pretty easy to get along with in terms of the requirements for repairing the road. Call it luck or just small-town politics, but compared to what the road commission could have required, I got off very easy. Before I accepted this job, I reviewed all of the requirements for making and repairing a road cut. If someone had enforced the written word on this issue to the limit, the road cut might have cost more than the price for the whole job.

The requirements for another road cut I was involved with were somewhat ambiguous and financially dangerous. The requirements went something like this: Any contractor making a road cut was responsible for public safety at all times while work was in progress. This could mean putting traffic cones or barrels out or paying people to guard the trench at all times. This would not have been a cheap venture.

Once the cut was complete and the trench was ready to be filled, requirements became very difficult to predict. For example, the requirements stated that a contractor was required to repair the road to state standards and satisfaction. It's possible to determine what the state standards are, but what's it going to take to satisfy someone? The requirements went on and on in this way. They basically left every loophole imaginable for the town or state to force more work out of a contractor. I can see a situation where a contractor under these circumstances might spend tens of thousands of dollars and still be liable to the authorities for additional work. If this doesn't scare you, I don't know what would.

The costs involved with cutting and repairing a paved road can reach astronomical numbers. Before you take on any job that will require tapping into a public water main, get all the facts pertaining to the work. Your plumbing costs will be predictable, but the road work might not be so clear.

Drilling for private water sources

If you work in an area where private water sources are more common than public ones, you will normally be working with wells. Drilled and dug wells are the most common types of private water sources, but they are not the only ones. Some people use driven wells to get their drinking water. There are even people who still get their water from natural springs. As a contractor, I would limit my offerings for private water sources to drilled and dug wells. Other types of private water sources are not as dependable and may not produce a good quantity or quality of water.

Drilling a drilled well

Drilled wells tend to be the most dependable type of well. The depths of these wells range from 100 to 500 feet. The diameter is small, usually about 6 inches (Fig. 11-1), but the depth of the well normally means that a strong water vein will be found. The production rates, which vary per well, can be 3½ gallons per minute, a good rating; 5 gallons per minute, very good; and above 5 gallons per minute, great.

What drilled wells lack in diameter, they make up for with height. Steel casing is used with drilled wells until bedrock is encountered. At this point, the rock becomes the casing. I have a drilled well that is a little more than 400 feet deep. The static water level in my well is only about 15 feet from the top edge of the casing. This means that my water reserve has a 6-inch diameter and a depth of about 385 feet. That's a lot of water!

Because drilled wells tap into water veins found deep in the earth, they are not as likely to dry up during dry spells. It is very rare for a

11-1
Steel casing and well cap for a drilled well

drilled well to run out of water. Unless something happens to alter the underground water way, such as the blasting of rock for a highway, a drilled well should produce good water in an adequate quantity.

Due to the depth of a drilled well, submersible pumps are normally installed to deliver water. A two-pipe jet pump can be used, but its performance and longevity is not as good as the submersible pump. Jet pumps are installed outside of wells, usually in basements or crawl spaces. Sometimes little pump houses are built for them near the well being pumped. Submersible pumps are suspended in the water held by a drilled well. The pump hangs from the pipe that delivers water up to the water-service pipe.

Digging a dug well

Dug wells can be dug by hand or with special well equipment. The latter is typically the case. These wells have large diameters, commonly a 3-foot diameter, with preformed, concrete-lined walls that are set in place, one on top of the other, to keep the well from caving in. The casing also prevents surface water from contaminating the well.

In terms of cost, dug wells are much more expensive than driven wells, but considerably less expensive than drilled wells. Due to the large diameter of a dug well, it is possible to maintain a significant reservoir of water. One disadvantage to dug wells, however, is that they have a tendency to run dry during periods of low rainfall.

The well shaft for dug wells, sometimes called shallow wells, typically do not exceed 40 feet; however, many are nowhere near that deep. Shallow-well pumps, also known as jet pumps, are normally used to deliver water from dug wells.

Pounding a driven well

Driven wells are fairly common in Maine. This is the only place where I have worked and found an abundance of driven wells. Personally, I don't like them. This type of well is not usually very deep, so I question the quality of the water. Due to the nature of driven wells, they do not normally produce a high volume of water. While they might work fine for a weekend cottage, I don't see how they can be dependable for a year-round home. There are people in Maine, and probably many other places, however, who use this narrow-diameter well as their only source of water.

What is a driven well? Most driven wells consist of a pointed end that serves as a filter, a shaft of pipe, and some type of pump arrangement. Driven wells, or points, as they are sometimes called, are very inexpensive to install, but the ground must be suitable for such an installation. Rocky ground, for example, does not allow the use of a driven well because the well pipe has to be driven into the ground. If rock is encountered, the installation is stopped. Basically, a point is attached to a steel pipe and a drive cap is placed on the other end of the pipe. The length of the pipe rarely exceeds 10 feet. Multiple sections are used as the point is driven into the earth.

Each time a section of pipe is driven into the ground to a position low enough to allow a new section to be added, a coupling is installed to accept another section of pipe. This process continues until water is found or until the pipe can no longer be driven into the ground. While inexpensive, the installation can be back-breaking work. Once suitable water has been found, it is pumped to a distribution point, usually located within the foundation of a building.

Driven wells have very small diameters. The only reserve water is what is standing in the drive pipe or pooling in the ground. The quantity of water available is usually very low. If a well of this type does not have a fast recovery rate, it is easy to pump the well dry, and this is not desirable.

Selecting a water pump

There are three basic types of water pumps used in modern plumbing systems: single-pipe jet pumps, two-pipe jet pumps, and submersible pumps. Single-pipe jet pumps are used when the requirements for lifting water are not very great. This would normally be the case with a dug well. These pumps are installed outside the well, and one pipe runs from the pump to the well.

When the lift requirements are more than a single-pipe jet pump can handle, you have two options: the two-pipe jet pump or the submersible pump. Both will pump water from deep wells, but submersible pumps are thought to be a better choice. Almost any plumber you ask will recommend a submersible pump over a two-pipe jet pump.

Pump selection is an important part of any job where water will be pumped. You first have to decide which type of pump to use. Most people opt for a single-pipe jet pump when working with a dug well and a submersible pump for deep wells. This part of the selection is pretty easy. Once you know which type of pump you want, it is necessary to size it properly.

To size a pump, you must know how far water will have to be lifted and the recovery, or production, rate of the well. If you install a pump that is capable of pumping 5 gallons of water a minute in a well that produces only 3 gallons of water a minute, you will have trouble. If the well is not capable of producing water as fast as the pump can pump it, you could run out of water.

As a general contractor, you should not have to assume responsibility for sizing pumps used on your job. This burden should rest on the shoulders of the subcontractor who is supplying and installing your pump. However, the more you know about pumps, the better off you are. Sizing data is available from pump suppliers and manufacturers, and I recommend that you solicit the sizing recommendations for the brands of pumps you use. This will allow you to double-check decisions made by your subcontractors.

Water pumps can become complicated when you have to size and troubleshoot them. Most of this work will be done by your subcontractors, but it helps to have a working knowledge of the equipment for which you are ultimately responsible. If you want to know more about water pumps, check out Chapter 14.

Adding a water service

What is a water service? Plumbers consider a water service to be the pipe that extends from a water source to a water-distribution system. Once a water service is extended 5 feet inside a foundation, it changes from a water-service pipe to a water-distribution pipe. This is important to note because the plumbing code allows some materials to be used for water services that are not allowed in a water-distribution system. This means that a plumber using a water-service-only pipe has to convert to a different type of pipe material within those 5 feet.

Water-service pipe is normally buried in the ground. The plumbing code requires a water service to be installed below the local frostline. Because the pipe is buried, it is best to avoid the use of couplings. This is not difficult to do with today's modern plastic pipe. Long rolls of this pipe are available, making it possible to run several hundred feet without a joint in the pipe.

A water-service pipe cannot lay side-by-side with a sewer line in a common trench. It must be laid on a shelf that is above the sewer line (Fig. 11-2). This shelf must be solid ground, so the person excavating the trench must be instructed to cut the shelf as needed. Installing the sewer and water service lines in a common trench reduces costs, but you must make sure that a shelf is provided.

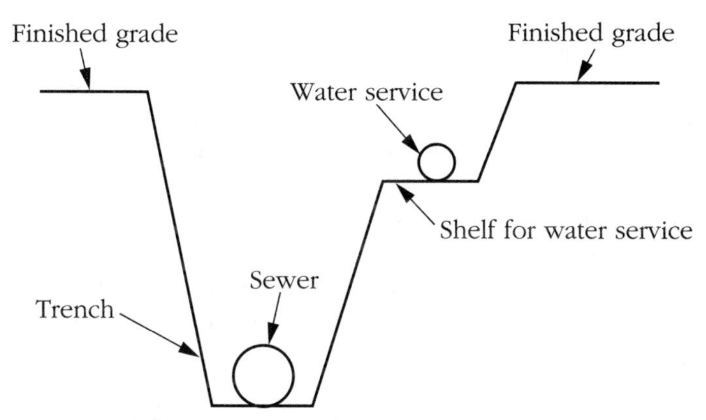

11-2 *Layout for installing a sewer and water service in a common trench*

Installing water-service materials

There are many types of water-service materials approved by the plumbing code; however, only three types of materials are commonly used. Copper tubing was king of the materials for a while, but this is no longer true. Modern plastic piping has taken over the top spot because it is less expensive than copper, and it is available in very long lengths, allowing long runs to be installed without couplings. This is a big advantage over copper tubing. Furthermore, plastic pipe is not subject to the same corrosion as copper. The ease of installation with plastic pipe gives it another leg up on copper. When all things are considered, plastic pipe is a much better choice for a water service than copper.

Using polyethylene pipe

Polyethylene (PE) pipe is the type of plastic pipe that is used most often for water services. It is a black pipe that comes in long coils. It can be difficult to work with in cold temperatures, but it is generally easy to install. This is also the type of pipe used most frequently during the installation of well pumps.

There are some drawbacks to PE pipe. One, it can kink quickly, especially if unskilled hands are installing it. Two, PE pipe becomes stiff in cold temperatures. Plumbers often have to warm the pipe to make joints in it during cold weather. This is not a big problem, but it is one potential downside to the pipe.

When joints are made with PE pipe, the process involves insert fittings and hose clamps. The insert fittings may be nylon, steel, or brass. Most plumbers you ask will agree that brass fittings are best. Nylon fittings tend to break if they are placed under stress. Experienced plumbers who are doing a good job use two hose clamps at each connection point. This is not a code requirement, but it is a good practice. The second clamp provides insurance that the joint will not come loose.

While cold weather causes PE pipe to stiffen, hot weather makes it soft. This can add to problems with kinking. It can also create some problems when the pipe is being covered up in a trench. If the backfilling is not done carefully, PE pipe can be crushed. This is especially true during hot spells.

A final drawback to PE pipe is that it is not rated to handle hot water. This makes it unsuitable for a water distribution pipe in a building where hot water will be available. Because of this, PE pipe may not extend more than 5 feet inside a foundation of a building where hot water will be used. This is not a major problem, but it does require a conversion to some other type of pipe for water distribution.

Choosing polybutylene pipe

Polybutylene (PB) pipe is my favorite. It is approved for hot-water usage, so it can be used as both a water service and as a water-distribution pipe. PB pipe is less prone to kinking than PE pipe and is extremely durable. Unlike PE pipe, PB pipe is normally joined with copper insert fittings and copper crimp rings. Due to the strength of PB pipe, hose clamps do not normally cinch the pipe tight enough to make a suitable connection. Compression fittings are sometimes used with PB pipe, but crimp connections are the most common method of making joints.

Like PE pipe, PB pipe is available in large coils, but it is also available in straight lengths, similar to rigid copper. When PB pipe is installed by competent professionals, it can give years and years of trouble-free service. As a builder and plumber, I wouldn't want to use any other type of pipe. There are occasions when I use PE pipe or copper tubing, but this is only if a customer makes a specific request. If the decision is left to me, I use PB pipe.

Backfilling the water-service line trench

Backfilling a trench where a water service is installed requires a careful approach. If you just take a back hoe and push large mounds of dirt and rock in on top of the water service, damage to the pipe is likely. A water-service line should be installed on solid ground that is free of any sharp objects, such as rocks. When the pipe is covered, it should done a little at a time. Dropping full buckets of dirt in a ditch can crush or kink a water service. Copper and PE pipe are more prone to this problem than PB pipe, but any of these materials can be damaged from reckless backfilling. To avoid having to dig up a damaged pipe, make sure that the person backfilling your trenches does so with care.

Water sources and water services are important parts of a plumbing system. The more you know about them, the better off you will be. Take some time to digest what you've learned, and remember to consider these factors in your next job. Not every job involves this type of work, but the ones that do deserve your attention.

12

Choosing sewer options and installations

Options and installations for sewers are vastly different, yet surprisingly similar to choices available for water sources and services. For example, the potential problems we discussed pertaining to road cuts in the last chapter also apply to sewer hookups. Tap fees are also involved when you put in a sewer. These aspects are not covered in this chapter because they were discussed in the last chapter.

As with water sources, both private and public sewer systems are in use. Municipal or city sewers offer the same types of advantages as city water mains. Private septic systems for sewage disposal are very different from water wells, but they can present similar difficulties for contractors.

Tapping into the city sewers

City sewers guarantee contractors a suitable means of sewage disposal. When a sewer main is available, a water main is usually available also. However, many communities offer public water without the option for tapping into a public sewer. If you are constructing in a major city or town, both are typically provided.

As a contractor, you have certain advantages when a city sewer is available. The biggest advantage is knowing that there is some place to drain the sewage from the building. Contractors in rural areas can't make this assumption. The tap fee charged to connect to a sewer can be substantial. Figures in excess of $2000 are not unusual, with amounts typically reaching much higher figures. If you are working up a price for building a new home or commercial building where a sewer tap is required, it is essential that you determine the tap fee before placing a bid.

Tap fees are not your only concern. Some sewers are very deep in the ground. A standard back hoe might not have enough reach to excavate a trench for your sewer. I ran into this problem on a townhouse project. Some of the sewer connections were nearly 20 feet below ground level. This type of digging requires a large excavator. If a contractor had planned to use a back hoe and discovered that an excavator was needed to accomplish the digging, the price of the job would skyrocket. This, of course, would result in lost profits and possibly lost out-of-pocket money. You have to establish your depth requirements before you can present an accurate bid.

Digging deep into the ground is not the only risk associated with municipal sewers. Road cuts can also eat into your profits quickly. We discussed road cuts in the last chapter, so I will not dwell on them here. Just be certain that you know what is expected of you if a road cut is needed.

Sewer laterals are often extended to a property line. When this is the case, making a sewer connection is easy. Having a lateral on-site takes the heavy financial risks off of your shoulders. If all your plumber has to do is tie into an on-site lateral, you don't have much to worry about. These connections are usually done with mechanical fittings, such as rubber couplings and hose clamps. The time required for a plumber to install a typical sewer from the foundation of a building to an on-site lateral is usually less than 4 hours, sometimes much less. This, however, does not include digging or filling the trench required for the sewer installation.

Installing private disposal systems

Private disposal systems for sewage are common in rural locations. These systems are most often referred to as septic systems. They typically involve a concrete septic tank (Fig. 12-1), a distribution box, and a network or pipe. Septic systems can be fairly simple in design and cost, but they can also become complex and expensive. Soil conditions are one of the biggest factors in determining the type of private sewage system needed.

Ground that has a good absorption rate will require a less-expensive septic system than ground that does not perk well. In ground where absorption is limited, the expense of a private disposal system can soar due to the use of chambers. When a septic system has to be installed higher than the drains feeding it, pump stations must be used. This obviously adds to the cost of a system.

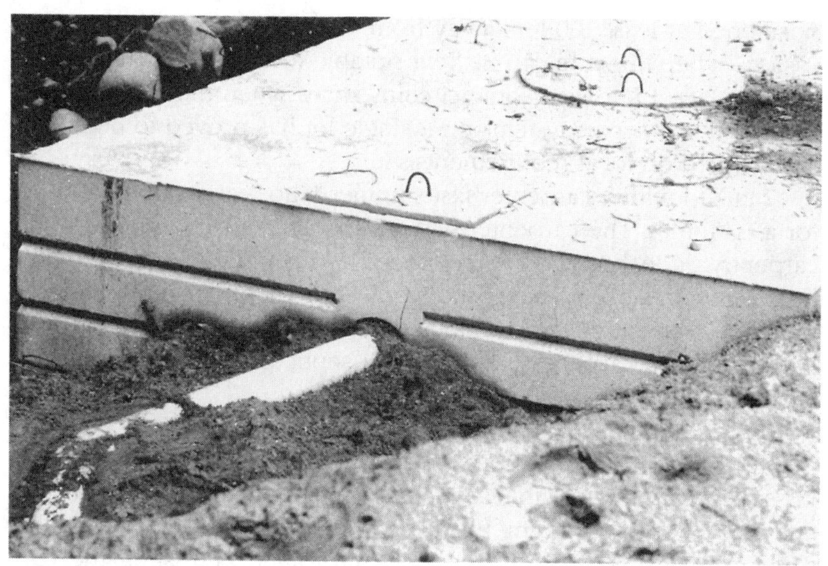

12-1 *Typical septic tank*

To what extent are you responsible for septic systems? The answer to this question depends on several factors. In my way of thinking, general contractors are responsible for every aspect of a job. You may contract a company to install a septic system and feel that your responsibility has been met. I don't share this feeling. For me, a job is not done until it is done, and I am the one person who has to answer to my customers. In other words, I feel that general contractors are responsible for all aspects of their jobs, and this includes septic systems.

Putting in a septic system

Who installs septic systems? My experience has shown that septic systems are most often installed by site contractors. Some companies specialize in septic systems, but they are rarely plumbing contractors. Generally, the septic installer runs a sewer to within 5 feet of the foundation of the building being served by the disposal system. Plumbers then make the connection between the building drain and sewer. Sometimes plumbers do run complete sewers to septic tanks, but they almost never install the tanks.

Choosing the septic system's location

Septic locations can become a sensitive issue, especially when a small building lot is being used, because it is hard to find a location for both a septic system and a well. Wells are generally required to be positioned at least 100 feet away from a septic system. There are occasions when this rule can make a suitable installation nearly impossible; in fact, I know of some circumstances when the installation of both well and septic systems on a single lot has proved to be impossible, based on local requirements.

I talked with a carpenter last summer who was building a house for a customer. The carpenter had several years of experience in the carpentry trade, but was inexperienced as a general contractor. When he took on the job, he never thought to get a septic permit before installing a foundation for the house. This proved to be a huge problem. When the carpenter applied for a septic permit, the application was rejected.

The building lot being used by the carpenter was small. Based on where the house foundation and the well were located, it was impossible to install a suitable septic system without violating local rules and regulations. According to the carpenter, his customer was rather upset. Understandably so, here's the customer, entrusting the construction of a home to a professional builder, only to have the builder create a problem that could not be solved without significant expense. You can imagine how the customer must have felt.

The carpenter came to me for advice. After reviewing the information given to me, the only alternative I could see that would not cost thousands of dollars involved getting an easement from an adjoining property owner. If such an easement was granted, part of the septic system could be installed on the adjoining lot. This would allow enough space between the well and the septic system to keep everything in order. Aside from a variance issued by local authorities, the only other solution involved moving a foundation that was already in place.

The carpenter and his customer talked to the adjoining landowner. They met resistance at first, but finally won the landowner over. I'm not sure of what financial arrangements were made, but I was told that permission was gained to place a portion of the septic system on the neighbor's property. While this proved to be the best solution to a bad problem, it still was not a great one. Having a portion of your septic system on your neighbor's property couldn't result in a real comfortable feeling. It was the better option, however, than the alternatives.

In the aforementioned example, the general contractor was at fault. A lawsuit could have developed easily over this situation. If the foundation had been moved, who do you suppose would have paid for the additional building costs? I suspect that the builder would have. This could have sucked all of the profits out of the job. This type of a problem is not common, but it can come up.

In my own building experience, I've had occasions when placing both a well and septic system on a lot has been a sticky situation. Fortunately, I've always discovered the potential for problems before any money was spent on construction. If I'm buying land on which to build, a septic design and well location are two of the first contingencies I place in a contract. I will not buy land until I'm sure that both water and sewer service is available. I recommend that you do the same. Buying a piece of property that is too small for a building, a well, and a septic tank can result in a lot of lost money.

Checking the soil characteristics

Soil characteristics play a big role in the size and design of a septic system. In some cases, land simply cannot be used to house a septic system. Some land can be fitted with a septic system, but at a cost much higher than anticipated. The following is an example of just such a case.

I was contacted by a couple who asked me to give them some quotes on a new house. The couple already owned the piece of land on which the house was to be built and had done most of the preliminary paperwork on the building themselves, such as getting individual estimates for all of the construction phases and acquiring a construction loan. Road work had also been started.

In my first meeting with the couple, we talked about septic systems. They told me that the estimates they had received were all around $4500. This price was based on a standard, gravel-bed, gravity-fed septic system for a three-bedroom house. The price they quoted seemed reasonable until I saw their building lot.

When I did a site inspection, I found the land to be very rocky. There was bedrock creeping up out of the land in the location chosen for the house. The only good soil I could find was up higher than the proposed house foundation. This alerted me to a potential problem.

After asking the customers about how they had gotten their prices, I was amazed to find that none of the contractors who were bidding the septic work made a site visit. Before I would proceed, I required the couple to obtain a septic design from a soils engineer. Sure enough, the only suitable location for the septic system was the one I had found. Because the land suitable for the system was high above the proposed foundation, a pump station would be needed. To add insult to injury, the perk rate on the site was minimal. This meant that a chambered system would have to be used rather than an economical gravel bed. When the dust settled, the new estimates on the septic system were in excess of $10,000. This was more than twice the amount budgeted for a septic system. As it turned out, the people were able to amend their financing and proceed with construction, but the results could have been much worse.

Take a moment to think about what I've just told you. The people who made the mistake in my example were average people. A professional contractor should have known better. Would you have known what to look out for? If you, as a general contractor, had made the same mistake, could you have afforded to lose $6000 on the job? I doubt it, so take an interest in all aspects of the work you are bidding or you could be in the same financial bind.

As a general contractor, it is often your responsibility to arrange a suitable site and design for a septic system. This is normally done in one of two ways. When I worked in Virginia, I would call a county representative to come out and do a soils analysis. In Maine, soils engineers are normally retained for this purpose. The big difference is in the cost. My cost in Virginia for an approved location and design was less than $50. In Maine, the price is generally around $250.

If a septic system will be required for a project you have bid on, get an expert to test the soil and design the systems. This is normally a part of the process for getting a building permit. Most jurisdictions will not issue a building permit until a suitable septic design is on file. The exception to this is, of course, when you can tap into a public sewer. If you don't know whom to contact for a septic design, check with your local code enforcement office. The people there can point you in the right direction.

Installing a gravel-bed septic system

Gravel-bed septic systems are the least-expensive type to install. This type of system involves a large concrete tank, usually with about a 1000-gallon capacity, a small distribution box, some pipe, and gravel. When this type of system can be used, the cost of a septic system can be half of what a chamber system costs.

When a septic design is drawn by experts, specifics for how a system must be installed will be given. Such specifics should include the size of the septic tank, either a 1000-gallon tank or a 1200-gallon tank; the number of drain lines needed; the required length of the lines; and the depth at which the lines must be buried.

Once you have an approved design, you can solicit bids from your subcontractors for installing the system. Septic installers cannot give you solid prices until they have an approved design from which to work. Don't accept ballpark prices—they will only get you in trouble. Never bid a job where a septic system is involved without getting rock-solid prices. The difference between a guesstimate and a quote could be $5000 or more.

Going with a chamber system

When soil is not suitable for a gravel-bed system, a chamber system might be used. This type of system is much more expensive. Chambers used in septic systems may be made of concrete or plastic. The purpose of the chambers is to hold the discharge from a septic system to allow a slower absorption rate. Slotted pipe is used in a gravel system, which requires the earth to be able to accept a rapid rate of delivery. When the ground doesn't perk well, chambers are used. Their use slows the discharge rate, which allows poorer ground to absorb the waste at a slower rate.

Using a holding tank

On some pieces of land, no septic system can be used because either the ground cannot absorb the waste from a septic tank or a drain field cannot be installed. One possible option under these conditions is the installation of a holding tank. This is basically a septic tank that is used strictly to collect waste and sewage. When the tank is full, a truck is called in to pump it out.

Typical septic systems allow effluent to discharge into a drain field. Regular septic systems still need to have sludge pumped out of them periodically, but the frequency of this pumping can be looked at in terms of years. When a holding tank is used, nothing is discharged. Routine pumping is necessary on a frequent basis. While this is far from ideal, it is better than having a piece of land that cannot be built on. Holding tanks may not be allowed in your area, so check with local authorities before assuming that you can use one.

Upgrading existing septic systems

Existing septic systems can pose problems for remodeling contractors. Very few remodelers consider septic systems when they are giving prices for adding living space to a home. Whether the new living space is in a basement, an attic, a garage, or is a new room addition, you might have a problem with the existing septic system.

The size of a septic system is determined, in part, by the number of bedrooms in a house. If a house has a septic system that was designed for three bedrooms, it might not be large enough to carry the load of a four-bedroom house. When a remodeler adds two bedrooms in the attic of a Cape Cod, a code violation might be occurring. This is a violation that the code enforcement office should catch before a permit is issued, but it doesn't always happen. Let me elaborate on the subject.

A builder I know built a three-bedroom house for a customer about two years ago. Part of the house deal included the construction of a large garage with a gambrel roof. Several months after the project was complete, the builder was called back to convert the upper level of the garage into living space. The converted space was to contain two bedrooms. The original septic system had been designed and approved for a three-bedroom house. By adding two bedrooms in the garage, there would be five bedrooms associated with the property and septic system.

The builder had already begun construction when a code officer came by the job and posted a Stop-Work Order. The reason given was the septic system. I'm not certain of all the details, but I was told that several meetings and arguments ensued over the work stoppage. In the meantime, the builder was not being paid for work that had been done. I guess it was quite a mess to sort out. In the end, an agreement was reached. More drain lines had to be added to the septic system before the job could be finished. The results of a problem like this could get quite out of hand, and you could have your own money tied up in a job that was stalled by a lawsuit. Don't fall into this type of trap.

Another type of risk is associated with existing septic systems and remodeling contractors. If you are digging footing for a new addition, you had better make sure that your excavation work is not near any existing drain lines. I've seen more than one job where contractors dug up drain lines by accident. This is a very expensive mistake to make. When you will be digging, confirm the location of all utilities and obstacles. Paying out of your own pocket to repair a septic system is not a good way to make money on a job.

Digging up drain lines is not the only way that you can put yourself in a financial bind with existing septic systems. Some drain fields are close to the top of the ground. Routing heavy trucks and equipment over a drain field can damage the field. Likewise, if drain pipes are crushed, they will have to be replaced. This is not only a messy job, but it is an expensive one. Septic tanks can also be damaged by trucks and equipment.

I don't have first-hand knowledge of crushing drain lines, but I can tell you about how I damaged a septic tank. When I first moved to Maine, I rented a house while looking for land on which to build. During the time I lived in the rental house, the landlord never told me where the septic tank or system was located. Based on the grounds around the house, I assumed the drain field was on the left side of the house under the only real lawn area. Ah, but I was wrong.

When the time came for me to move out of the rental house, I drove a large moving truck into the driveway. I pulled the front of the truck into the edge of some woods so that I could back it up to the door for easy loading. As the truck rolled into the leaves of the wooded area, the front wheel on the passenger side dropped down considerably. The big truck pulled itself out of the rut easily, so I thought nothing more of it. I just assumed that I had dipped into an old stump hole, but I was wrong.

The next day, when I was back at the house with my pick-up truck, the landlord approached me with a grim look on his face. He informed me that I had ruined his septic tank. I was shocked to hear this. How could a septic tank be buried in the woods? Tree roots and drain fields just don't go together. Sure enough, though, I had put the front tire of the truck on the top of an old metal septic tank. The tank was so rusted that the truck's weight crumpled its upper lip. Now it would have been bad enough to pay for a new septic tank, but the circumstances were much worse. You see, the old, 500-gallon metal tank was legal under a grandfather clause, but it could not be replaced. The tank and the field were far too small to meet modern code requirements. And yes, the field was under the trees, if there was a field at all. I could see a horrible lawsuit coming my way.

Due to this house's small lot, there was nowhere to install a legal septic system under modern requirements. By driving the moving truck over a septic tank that I had no idea even existed, I'd set myself up for a steep financial fall. Fortunately, I was able to bring in a septic expert to repair the tank. Repairs were acceptable under current code requirements, but replacement wasn't. I dodged a very expensive bullet, but it left me with a lasting memory.

Before you put trucks or equipment on a remodeling site, make sure you are not going to damage anything that you can't see. It might be best to require a written waiver of liability from the property owner, just to be safe. The cost of some damages can be enough to put you out of business.

13

Punching up your profits with plumbing fixtures

How would you like to punch up your profits with plumbing fixtures? You can increase your profits with plumbing fixtures by supplying the fixtures yourself, selling your customers a better grade of fixtures, and making sure your fixtures are being productively installed. Fixtures account for a large percentage of plumbing costs for a job. If you learn to use this phase of plumbing to your advantage, you can make more money, and I'm going to show you how.

Marking up your plumbing fixtures

The markup on plumbing fixtures can be quite handsome. For example, some plumbers buy toilets for less than $60 and sell them for $125. Faucets for which a plumber pays $8 might sell for $25. Better grades of faucets might be sold to a plumber for $50 and to a customer for $115. There is a lot of money to be made in the markup of materials. Of course, you already have some idea of how marking up materials can benefit you, but the percentage of markup on plumbing fixtures is often much greater than what can be earned with lumber and basic building materials. For example, when was the last time that you could charge $5 for a stud that cost you $2.50?

As you move into big-ticket items, like whirlpool tubs, the money made on the markup can become quite substantial. For instance, a whirlpool that a plumber pays $750 for might well retail for $1500.

Give a little thought to this. If you are a remodeling contractor who is redoing a bathroom, consider how much extra money you could make by providing the plumbing fixtures yourself. You might pick up $750 for a whirlpool, $200 on faucets, and $75 on a toilet. This is more than $1000 made just for arranging delivery of the fixtures. Now compare the $1000 made from the fixtures with your total profit on the job. I think you will see that the material markup comprises a large percentage of your overall profit.

If you were remodeling or adding a bathroom with first-class fixtures, your profit from the sale of the fixtures could be considerably more. For instance, a pedestal lavatory in a designer color can produce a high rate of return. Anytime high-fashion colors are used, you can count on big money from the sale. If your customers like the finer things in life, such as gold faucets, you'd be amazed at how much money you can make just from being a dealer for the faucets.

Are there people out there who will pay thousands of dollars for one faucet? Yes, there definitely are. I was selling gold faucets with some regularity 12 years ago. High-fashion fixtures are popular among people in certain circles. If you get into the right niche, you can walk away from a small remodeling job with thousands of dollars in profit, derived solely from the markup on fixtures. This may seem a little hard to believe, but I know it's true—I've done it.

Plumbers like to provide fixtures and rough materials for the jobs they do because they can make more money that way. As a general contractor, you get to call the shots. If you insist on providing your own materials for plumbing work, you're going to find some plumbers who will accept your terms, others who will not be happy giving up the easy money but will still want your work, and some who may flat out refuse to take on work when they don't supply materials. The following sections highlight the obvious reasons why plumbers like to supply their own fixtures.

Losing money

One of the main reasons why plumbers resist working for contractors who want to supply all of the plumbing materials is because they lose money in the process. Any plumbing contractor who has been in business for a year or so knows the economic value of providing materials. They also know the risks involved with providing materials. When plumbers pay for materials out of their own pockets and supply them, they are at risk. If you, the general contractor, refuse or fail to pay the plumbers, they are out of more than just their labor. They lose the money they invested in materials. This situation gives you a bargaining chip to use with your plumbers.

The resentment that some plumbers harbor for contractors who want to cut them out of the markup profit can reach high levels. When a contractor informs plumbers that they are not allowed to bid anything but labor on a job, there is a chance that the plumbers will refuse to show any interest in the work. There are usually plenty of plumbers who will bid labor-only jobs, but some of the most desirable plumbing contractors may not. As a general contractor, you have to find a way to soften your position. This doesn't mean handing all the material profits over to your plumbing contractors, it only means that you must present your case in a better way.

One of the most effective ways for a general contractor to get the cooperation from plumbers on the issue of supplying materials is to point out the advantages of having the general supply the materials. Plumbers who are not allowed to supply fixtures do stand to lose some profit, but they are not at risk of losing money they have already made. If they supply fixtures and are never paid for them, they lose profits and out-of-pocket money. A savvy contractor can use this fact to sway plumbers into a more agreeable position.

If you explain to your plumbers that you are taking all of the risk for getting paid for the supplied fixtures, you might see a more receptive audience. Pointing out that a plumber who invests a few thousands dollars in fixtures may never see the money again is a good way to drive your point home. If you manage this with tact, your plumbers will feel as though you are protecting them. This makes you a good guy. It's all in the presentation.

Having to wait for materials

Some plumbers don't want anyone else to supply their materials out of a fear of lost time. When plumbers need specific items to maintain maximum production, they can't afford to stand around waiting for the materials to become available. This concern often weighs heavier on the minds of plumbers than the loss of potential profits.

Some contractors are not very good in their field management. If I were your plumbing contractor and you sent me out on jobs where needed materials were not available, I wouldn't work with you for long. Time is a very important commodity to a plumber. Lost time translates into lost money. If you are going to supply materials for your plumbers, you must make them available when they are needed. Failure to accomplish this goal will result in a high turnover of plumbing contractors.

Ordering the wrong materials

Having the wrong materials on a job can ruin your plumber's productivity. For example, if you have a bathtub that is cracked on a job, the plumber is going to lose time while you reorder the tub. Likewise, if you order 4-inch-center faucets for a lavatory that requires 8-inch-center faucets, your plumber is going to have to make an extra trip to finish the job. This means that the plumber will lose time and money unless you are back-charged for your mistake. This type of situation has to be taken into account before you make a commitment to supply the materials. If you can't accurately predict the needs of your plumbers, you might be better off to let the plumbers provide their own materials.

Making extra money

Money is usually the big issue when it comes to plumbing fixtures. Plumbers want to supply them to make extra money. General contractors want to supply the fixtures to put more cash in their pockets. Sometimes a compromise is needed. One plan that should work well for both parties is to have your plumbing contractors supply all rough material while you supply all fixtures. This eliminates a lot of the risk that plumbers take in running out of needed material during a rough-in.

It is fairly easy for a general contractor to predict what types of plumbing fixtures will be needed for a job. For instance, the contractor should know how many faucets, toilets, lavatories, sinks, and similar fixtures are needed. It is not so easy, however, for someone who is not a plumber to determine what types of rough materials are needed and what quantities are needed. You can count up the number of angle stops needed and figure out how many traps are required, but when it comes to pipe and fittings, most general contractors cannot accurately predict the plumber's needs.

I have worked on houses in the past where plumbing material was included as part of the overall material package for building the dwelling. I have never done a prepackaged job where all the required pipe and fittings were supplied. Part of this is due to code variations. A plumbing package that might meet code on the West Coast may not do so on the East Coast. Because packaged houses are not created on a code-by-code basis, certain mistakes in material take-offs are likely. The times are rare when someone other than a plumber can project the exact needs for a job.

I worked for a company a long time ago that predetermined, with the help of computers, the exact type and number of fittings needed to plumb each house in a housing project. The townhouses where this procedure was used made it possible for the estimates to be very close to perfect. However, even though every sixth house was essentially the same, the computerized list of materials often failed. If carpenters framed a townhouse a little differently, a new selection of fittings was required. It is very frustrating for me, the employee plumber, to need one specific fitting to complete a job and not have it. The only thing that would have made it worse would have been if I had been an independent contractor. As an employee, I got paid to wait for additional fittings. If I had been in business for myself at the time, I would have been losing money.

The problem that must be addressed is one of money. Your plumbers want to make it and so do you. If you have a good plumbing contractor, you don't want to jeopardize the working relationship to save $100 here and there. It just isn't worth it. It is foolish, though, for you to give up profit dollars that could be deposited in your bank account. The decision is yours.

Setting a limit on your markup

How much markup can you place on an item without upsetting your customers? This a question that does not have a uniform answer. Some plumbing products can be marked up to an outrageous profit percentage without risking much reprisal from customers. Other types of plumbing materials don't allow for much of a markup.

Can you buy materials from plumbing contractors and mark them up to your customers? Yes, most general contractors do. It is not at all uncommon for general contractors to add 10% onto the costs of a job. Some generals work with a 20% markup. If general contractors didn't do this, they wouldn't make much money. But how far is too far?

If you get greedy, your customers may pay your price once but never call you again. It is senseless to alienate customers by gouging them with your prices. This, however, doesn't mean that you are not entitled to a fair markup. With plumbing fixtures, a fair markup can be quite a few dollars.

There are different schools of thought on how to determine the markup on materials. Some contractors keep their markup percentages low to make their bids more attractive. If they win a job in this way, they are happy to get the work. A few contractors stick unreasonably high prices on their goods and services and wait. Sooner or later, they wind up getting some percentage of the jobs they bid. The ideal place to be is somewhere between these two extremes.

Pricing your plumbing fixtures

Prices for plumbing fixtures and materials vary tremendously. Most customers judge a contractor by who has the lowest bid for fixtures and materials. This is true for most large jobs where it is difficult for consumers to put individual price tags on objects, but if you are doing small jobs, in terms of materials, it is easier for customers to compare your prices to the ones in the stores. When you are charging $1500 for a whirlpool tub that is available to the public in stores at a price of $1200, your customers might be offended. In fact, they may get downright angry. When a customer can pinpoint individual prices in a bid, you must make sure the prices you are quoting are fair.

I've had customers talk to me about other plumbing contractors who they feel took advantage of them. For example, I remember one homeowner who showed me a fairly new kitchen faucet and asked me how much I would charge for a faucet of its type and brand. When I told the customer what my price would be, it was as if I had stepped on her toe. Her mouth fell open and her face showed pain. After this initial reaction, she told me what a competitor of mine had charged for the faucet. Then I could see why she was upset. The other plumber had sold the faucet at a price that was considerably higher than the highest retail price I had ever seen for it. This type of situation costs you customers. You might make a little extra money at the time of your sale, but the results are likely to hurt you in the long run.

I never price my fixtures and materials higher than what the local stores are offering for the goods. If a customer does a direct price comparison, I'm in line with other retailers. This eliminates any bad feelings between my customers and my company.

Some contractors use a fixed markup percentage to determine their sales price. I don't. By monitoring the local market, I charge competitive prices for what I sell. This gives me a 20% markup on some items and a 90% markup on others. This procedure allows me to make the most money possible without risking the relationships I build with my customers. Let me give you some examples of various material prices in my local region.

Billing toilets

Toilets are a common plumbing fixture. As a plumber, I can buy toilets for less than $50, but I generally spend around $70. This is because I use a base grade that is slightly higher than a basic builder grade. Some local stores retail toilets to customers in my area at prices of about $80. It is not uncommon for better grades of toilets to sell for $125. Most customers don't know a lot about toilets, so I could sell a $50 toilet for $125. This would certainly be a good markup to make, but I suspect a customer would call me on the price at some point. If I can make $55 in profit by selling a toilet, I'll take it, but there is no reason for me to go out on a limb just to make an extra 20 bucks.

Charging for faucets

Some of the low-end faucets I offer customers cost me less than $10. I can easily double the price of what I pay for these faucets without raising any problems with my customers. A 100% markup is very good from a numbers point of view. Better faucets, like the ones I normally use, cost in the neighborhood of $50. These faucets are passed on to my customers at prices of about $75. My percentage of markup is much lower on these faucets, but the profit is still good.

Selling the big-ticket items

Big-ticket items, like whirlpool tubs, can offer tremendous profit power. With some brands of whirlpools, it is possible to double your money, but most won't allow such a large markup. Even at conservative markup percentages, though, big-ticket items will net several hundred dollars for the seller.

When you are setting prices, you have to look at more than just a fixed percentage of markup. While you can double the cost of a $10 faucet, doing the same thing with a $1500 whirlpool will normally put your price well above the normal retail value. People aren't stupid. They will go around looking at products in stores. If they find that you have taken advantage of them, you will lose not only that customer, but potential customers who have been told about your practices. It is not necessary to price your materials so high that consumers turn against you. In fact, it's just plain bad for business.

Cutting out the middle man

We all know that cutting out the middle man in a transaction is one way to save money. In effect, that is what we are talking about in this chapter. As a general contractor, you are cutting your plumber (the middle man) out of the profit picture as it relates to plumbing fixtures. Doing this will save you money on paper, but it might cost more than it's worth.

As I said earlier, many plumbers will work for contractors who insist on supplying materials, but some won't. If you have a good plumbing contractor who is accustomed to supplying his or her own materials, you may cause a lot of trouble for yourself by taking over the supplying process. Good plumbers are hard to come by, so breaking off a solid working relationship with your plumbing contractor could be much more costly than the money you would make by supplying the materials.

Assuming that your plumbers are okay with you being the supplier, you've got to establish yourself as a wholesale buyer of plumbing materials. This may not be easy to do. Some plumbing suppliers will only sell to licensed plumbers. Before you take a chance on upsetting your plumber, make sure that you can gain access to wholesale plumbing prices.

Many local plumbing suppliers will not sell to the general public. Some of them will not sell to anyone other than licensed plumbers. Being a licensed general contractor may not make any difference to some suppliers where you want to do business. This is not always the case, but you should be aware of the potential problem before you put too many wheels in motion.

Many building supply stores sell plumbing materials, but they usually charge a lot more for them than standard plumbing wholesalers. It is often possible to buy some plumbing fixtures from building suppliers at good prices. If you can do this, you're all set. What happens, though, if your local plumbing suppliers won't deal with you and your building suppliers don't have what you need or won't give it to you at a good price? What are your options?

I deal with one major plumbing supplier by mail. The company sends me catalogs on a regular basis and I buy a majority of my supplies from them. Why do I buy through a mail-order catalog? It's a lot cheaper to buy items by mail order than to buy from my local suppliers. The company I deal with pays the shipping costs for any order over $600 and the prices I pay range up to 30% less than what I would pay for an identical item at my local wholesaler. I get name-brand material for much lower prices, and they are delivered right to my door. There is never any standing in line at a supply counter; I just phone or fax in my order, and in a day or two, it arrives. I simply can't beat the price or service. If you have trouble making deals on your local level, look to suppliers outside of your immediate work region. Someone will be happy to sell to you at professional prices.

Selling up

In sales, there is a term called selling up. This simply means that you sell a customer a more expensive item. Even if you don't want to be bothered with supplying plumbing fixtures yourself, you can still increase your profits by selling your customers up. Let me give you an example of how this might work.

Assume you have a customer who wants a whirlpool tub installed. The customer has decided on a four-jet tub from a no-name manufacturer. We'll say that the price to you for this fixture is $1200. In this case, let's assume that you are allowing your plumbing contractor to supply all materials, but that you are adding a 20% markup to the prices given to you by subcontractors. This means that you will make $300 on this whirlpool because you are selling it to the customer for $1500. There is nothing wrong with this, but you might be able to do better.

Suppose you were able to sell your customer up to a six-jet, name-brand tub. What would your profit be then? For the sake of the example, let's say that the price from your plumber on this upgraded whirlpool is $1800. Assuming the same percentage of markup, the price to the customer would be $2250. Your profit has jumped from $300 to $450. In percentages, that's one heck of a big jump, and it's not bad in actual dollars. By selling customers up, you can increase your profits without doing any additional work beyond making the sale. This is a sweet way to build up your company's financial statement.

The principles involved with selling up can be applied to almost any plumbing fixture. If a customer is thinking of buying plain white fixtures, get them interested in buying high-fashion colors. The cost difference is a lot, so your profits will grow. Talk a customer into buying a one-piece toilet, instead of a dull two-piece unit. Replace an old wall-hung lavatory with a pedestal lavatory. Push the advantages of anti-scald tub-and-shower valves. If you work on your sales ability a little, you can create more cash with every job you sell.

Increasing productivity

Productivity and profits go hand in hand. If your plumbers can hit a job and knock it out fast, they can charge you less than what they would charge for a job that went slowly. Being a general contractor, you can control when your plumbers come in and make sure that your jobs are ready for plumbers when they arrive. If you impress upon your plumbing contractors how efficiently your jobs are managed, they may reduce their prices for you on future jobs.

Setting fixtures can be a frustrating job for plumbers. Being a plumber, I know all about the stress of being on a job that is not ready. For instance, getting to a new house to set fixtures only to find that none of the counters have been installed, which means that sinks and lavatories can't be installed, is an instant headache. Likewise, having to make a second trip to a job to set a toilet that couldn't be set with the rest of the fixtures because the bathroom floor wasn't ready is equally frustrating.

Plumbers who are good are busy. When plumbers are busy, they can't afford to waste time with extra trips back to jobs that should have been finished. If you build a reputation for yourself as a contractor who cries wolf in terms of being ready for your plumbers, word will get around. Plumbers will still work for you, but they will probably allow for your shortcomings as a general contractor by tacking extra money onto their bid prices. Conversely, if you become known as a well-organized general who never calls plumbers before a job is ready, you will have a flock of plumbers wanting to do your work. This is something that you should strive to achieve.

Troubleshooting fixtures

As a general contractor, you may find it helpful to know how to troubleshoot plumbing fixtures from time to time. To make this task easier, I've provided numerous troubleshooting tables for your use (Tables 13-1 through 13-8).

Table 13-1. Toilet troubleshooting

Symptoms	Probable cause
Will not flush	No water in tank
	Stoppage in drainage system
Flushes poorly	Clogged flush holes
	Flapper or tank ball is not staying open long enough
	Not enough water in tank
	Partial drain blockage
	Defective handle
	Bad connection between handle and flush valve
	Vent is clogged

Symptoms	Probable cause
Water droplets covering tank	Condensation
Tank fills slowly	Defective ballcock Obstructed supply pipe Low water pressure Partially closed valve Partially frozen pipe
Make unusual noises when flushed	Defective ballcock
Water runs constantly	Bad flapper or tank ball Bad ballcock Float rod needs to be adjusted Float is filled with water Ballcock needs to be adjusted Pitted flush valve Undiscovered leak Cracked overflow tube
Water seeps from base of toilet	Bad wax ring Cracked toilet bowl
Water dripping from tank	Condensation Bad tank-to-bowl gasket Bad tank-to-bowl bolts Cracked tank Flush-valve nut is loose
No water comes into the tank	Closed valve Defective ballcock Frozen pipe Broken pipe

Table 13-2. Kitchen sink troubleshooting

Symptoms	Probable cause
Faucet drips from spout	Bad washers or cartridge Bad faucet seats
Faucet leaks at base of spout	Bad "O" ring
Faucet will not shut off	Bad washers or cartridge Bad faucet seats
Poor water pressure	Partially closed valve Clogged aerator

Table 13-2. Continued

Symptoms	Probable cause
	Not enough water pressure Blockage in the faucet Partially frozen pipe
No water	Closed valve Broken pipe Frozen pipe
Drains slowly	Partial obstruction in drain or trap
Will not drain	Blocked drain or trap
Gurgles as it drains	Partial drain blockage Partial blockage in the vent
Won't hold water	Bad basket strainer Bad putty seal on drain
Spray attachment will not spray	Clogged holes in spray head Kinked spray hose
Spray attachment will not cut off	Bad spray head

Table 13-3. Lavatory troubleshooting

Symptoms	Probable cause
Faucet drips from spout	Bad washers or cartridge Bad faucet seats
Faucet leaks at base of spout	Bad "O" ring
Faucet will not shut off	Bad washers or cartridge Bad faucet seats
Poor water pressure	Partially closed valve Clogged aerator Not enough water pressure Blockage in the faucet Partially frozen pipe
No water	Closed valve Broken pipe Frozen pipe
Drains slowly	Hair on pop-up assembly Partial obstruction in drain or trap Pop-up needs to be adjusted

Symptoms	Probable cause
Will not drain	Blocked drain or trap Pop-up is defective
Gurgles as it drains	Partial drain blockage Partial blockage in the vent
Won't hold water	Pop-up needs to be adjusted Bad putty seal on drain

Table 13-4. Laundry tub troubleshooting

Symptoms	Probable cause
Faucet drips from spout	Bad washers or cartridge Bad faucet seats
Faucet leaks at base of spout	Bad "O" ring
Faucet will not shut off	Bad washer or cartridge Bad faucet seats
Poor water pressure	Partially closed valve Clogged aerator Not enough water pressure Blockage in the faucet Partially frozen pipe
No water	Closed valve Broken pipe Frozen pipe
Drains slowly	Partial obstruction in drain or trap
Will not drain	Blocked drain or trap
Gurgles as it drains	Partial drain blockage Partial blockage in the vent
Won't hold water	Bad basket strainer Bad putty seal on drain

Table 13-5. Shower troubleshooting

Symptoms	Probable cause
Won't drain	Clogged drain Clogged strainer Clogged trap

Table 13-5. Continued

Symptoms	Probable cause
Drains slowly	Hair in strainer Partial drain blockage
Gurgles as it drains	Partial drain blockage Partial blockage in the vent
Water drips from shower head	Bad faucet washers or cartridge Bad faucet seats
Faucet will not shut off	Bad washers or cartridge Bad faucet seats
Poor water pressure	Partially closed valve Not enough water pressure Blockage in the faucet Partially frozen pipe
No water	Closed valve Broken pipe Frozen pipe

Table 13-6. Bathtub troubleshooting

Symptoms	Probable cause
Won't drain	Clogged drain Clogged tub waste Clogged trap
Drains slowly	Hair in tub waste Partial drain blockage
Won't hold water	Tub waste needs to be adjusted
Won't release water	Tub waste needs to be adjusted
Gurgles as it drains	Partial drain blockage Partial blockage in the vent
Water drips from spout	Bad faucet washers or cartridge Bad faucet seats
Waters comes out spout and shower at the same time	Bad diverter washer Bad diverter seat Bad diverter
Faucet will not shut off	Bad washers or cartridge Bad faucet seats

Symptoms	Probable cause
Poor water pressure	Partially closed valve Not enough water pressure Blockage in the faucet Partially frozen pipe
No water	Closed valve Broken pipe Frozen pipe

Table 13-7. Electric water heater troubleshooting

Symptoms	Probable cause
Relief valve leaks slowly	Bad relief valve
Relief valve blows off periodically	High water temperature High pressure in tank Bad relief valve
No hot water	Electrical power is off Elements are bad Defective thermostat Inlet valve is closed
Too little hot water	An element is bad Bad thermostat Thermostat needs to be adjusted
Too much hot water	Thermostat needs to be adjusted Controls are defective
Water leaks from tank	Hole in tank Rusted-out fitting in tank

Table 13-8. Gas water heater troubleshooting

Symptoms	Probable cause
Relief valve leaks slowly	Bad relief valve
Relief valve blows off periodically	High water temperature High pressure in tank Bad relief valve
No hot water	Out of gas Pilot light is out Bad thermostat

Table 13-8. Continued

Symptoms	Probable cause
	Control valve is off
	Gas valve closed
Too little hot water	Bad thermostat
	Thermostat needs to be adjusted
Too much hot water	Thermostat needs to be adjusted
	Controls are defective
	Burner will not shut off
Water leaks from tank	Hole in tank
	Rusted-out fitting in tank

14

Troubleshooting well pumps

Builders who work in rural communities must often depend on well pumps as a water supply. If a pump fails, customers will start complaining very quickly. When a pump of the wrong type or size is installed, the general contractor often has to take the heat for it, even if it is not the general's fault.

How much do you know about water pumps? Do you know why a one-pipe jet pump is limited on the height to which it can pump water? Why do deep-well jet pumps have two pipes? What are the advantages offered with submersible pumps? If you can't answer these questions, you should take some time to study this chapter closely. Customers might not expect you to install their well pump, but they will want you to be able to answer their questions. Because the concerns of consumers can encompass a lot of areas, a general contractor must be prepared to answer the customer's questions or put all fears to rest about well pumps.

Unless you install pumps for a living, you don't have to know all of the technical facts pertaining to them. Whether a submersible pump is a two-wire version or a three-wire version may not matter to you. It may be your philosophy to pass all pump questions on to your plumber or well driller. Deferring questions to experts is not a bad idea, but there are times when pump experts are not available when you really need them. If you're sitting down with customers and on the verge of closing a sale, it would be a shame to lose the magic of the moment by having to wait until the next day to get answers on a pump question. This type of problem does come up.

The general public sometimes knows more about trade issues than the contractors in the trade. I don't mean that they are better qualified to be a general contractor or a plumber, but they may have studied an issue to a point where they have better knowledge of it. A customer who has been looking into the pros and cons of a two-pipe jet pump and a submersible pump may know a great deal more about their differences than an average general contractor. I don't think any reasonable person expects a general contractor to have an answer for every question, but if you can't answer a majority of them, you may lose some credibility and, possibly, the job.

I'm not going to attempt to turn you into an expert pump installer or troubleshooter; that would require extensive field work. I do, however, plan to give you some pointers that may come in handy when selling jobs and fielding customer complaints. The information presented to you in this chapter can also prepare you for dealing with subcontractors who will be installing and troubleshooting your pump systems.

I'm not going to waste a lot of time going back over the different types of pumps. You can refer back to Chapter 11 for this information. With the information in this chapter on the installation of water pumps and some rule-of-thumb troubleshooting tips, you should be well out in front of your competitors if a pump question arises.

Choosing and installing a jet pump

There are two types of jet pumps: single-pipe jet pumps, for use in shallow wells, and two-pipe jet pumps, for deep wells. Because single-pipe jet pumps are suction pumps, their use is limited to installations where the maximum lift requirement is less than 30 feet. With a perfect vacuum at sea level, a shallow well pump might be able to lift water to 30 feet, but maximum lift is not recommended and rarely achieved. Single-pipe jet pumps suck water up the well pipe and into a water-distribution system.

The two-pipe jet pump is sometimes called a deep-well jet pump. These pumps resemble shallow-well pumps in that they are also installed above ground. The operating principles of the two types of jet pumps differ. While shallow-well pumps suck water up from the well, deep-well jet pumps push water down one pipe and suck water up the other. This is why there are two pipes on deep-well jet pumps. The only major installation differences between a shallow-well and a deep-well pump is the number of pipes used in the installation and a pressure control valve.

The basic installation of a jet pump is not difficult to understand. While it is unlikely that you will be installing your own pumps, it is helpful to have some idea of what is involved with an installation.

When you install a shallow-well jet pump, the single pipe from the well to the pump usually has a diameter of 1¼ inches. A standard well pipe material is polyethylene, rated for 160 psi. You want to be sure the suction pipe is not coiled or in any condition that might cause it to trap air. If the pipe holds an air pocket, priming the pump can be quite difficult. In most cases, a foot valve is installed on the end of the pipe that is submerged in the well.

Once the drop pipe (the pipe going down into the well water) reaches the upper portion of the well, it usually takes a 90-degree turn to exit the well casing. This turn is made with an insert-type elbow. Always insist on the use of two clamps to hold the pipe to its fittings. When the pipe leaves the well, it should be buried underground and treated as a water service.

Once inside the foundation of a building, you might wish to convert to some other type of pipe. The water supply pipe should run directly to the pump. The foot valve that is in the well on the end of the drop pipe acts as a strainer and a check valve. When you have a foot valve in the well, you don't need a check valve at the pump. When the supply pipe reaches the pump, you can follow instructions provided by the pump manufacturer for the remainder of the installation.

Deep-well pumps and shallow-well pumps both use a foot valve. There is a jet-body fitting that is submerged in the well and attached to both pipes and the foot valve when a deep-well pump is used. The pressure pipe connects to the jet assembly first. The foot valve hangs below the pressure pipe. The suction line connects to a molded fitting on the jet body. With this jet body, both pipes are allowed to connect in a natural and efficient manner.

Deep-well jet pumps push pressure down the pressure pipe with water pushed through the jet assembly. This makes it possible for the suction pipe to pull water up from a deep well. From the suction pipe, water is brought into the pump and distributed to the potable water system. When you look at the head of a deep-well jet pump, you will see two openings where your pipes can be connected. The larger opening is for the suction pipe and the smaller one is for the pressure pipe. The suction pipe usually has a diameter of 1¼ inches. The pressure pipe typically has a diameter of 1 inch. Aside from the extra pipe and jet assembly, deep-well jet pumps are about the same as shallow-well jet pumps.

Selecting and installing submersible pumps

Submersible pumps are installed down the well shaft, submerged under the water at the bottom. This type of pump has only one pipe, through which water is pushed. Submersible pumps are much more efficient and easier to install than jet pumps. Under the same conditions, a half-horsepower submersible pump can produce nearly 300 gallons more water than a half-horsepower jet pump. With so many advantages, it is almost foolish to use a jet pump when you could use a submersible pump.

Installing a submersible pump requires a different technique than that used to install a jet pump. Because submersible pumps are installed in the well, electrical wires must be run down the well to the pump. Before you install a submersible pump, consult a licensed electrician for the wiring needs of your pump. You need a hole in the well casing to install a pitless adapter. The pitless adapter provides a watertight seal in the well casing for your well pipe to feed your water service. When you purchase your pitless adapter, it should be packaged with instructions on what size hole is needed in the well casing.

You can cut a hole in the well casing with a cutting torch or a hole saw. I prefer using a hole saw. The pitless adapter attaches to the well casing and seals the hole. On the inside of the well casing, you have a tee fitting on the pitless adapter. This is where your well pipe is attached. The tee fitting is designed to allow you to make all of your pump and pipe connections above ground. After all the connections are made, you lower the pump and pipe into the well and the tee fitting slides into a groove on the pitless adapter. A length of steel threaded pipe, about 8 feet long, is used to guide the tee fitting into its holder.

To make up the pump and pipe connections, you need to know the depth of the well. The well driller should provide you with the depth and rate of recovery for the well. Once you know the depth, cut a piece of plastic well pipe to the desired length. The pump should hang at least 10 feet above the bottom of the well and at least 10 feet below the lowest expected water level.

A male-insert adapter is screwed into the pump. A torque arrester is placed over the end of the pipe. The torque arrester absorbs thrust and vibrations from the pump and helps to keep the pump centered in the casing. Two stainless steel clamps should be use to secure the pipe to the insert fitting (Fig. 14-1). The tee fitting, or clamp, for the pitless adapter is installed on the end of the pipe at the top of the well casing.

Torque stops are installed on the pipe at regular intervals to prevent the pipe and wires from scraping against the casing during operation. Electrical wiring is secured to the well pipe at regular intervals to eliminate slack in the wires. Once all this has been done, the pump is ready to be lowered.

Before lowering the pump, it is a good idea to tie a safety rope onto the pump. The rope is tied at the top of the casing to prevent

14-1
Insert fitting

the pump from being lost in the well if the pipe and pump should separate. Not all pump installers do this, but you should insist on it.

When the entire piping arrangement is put together, the pump and pipe are lowered into the well. Caution must be used to avoid damaging the insulation around the electrical wires. Once the tee fitting is in position, it is fitted into the pitless adapter and the steel pipe used to lower the assembly is removed. At this point, a water service is connected to the pitless adapter on the outside of the well casing and run to the foundation where the water-distribution system is located.

Using pressure tanks

All well pumps should work in conjunction with a pressure tank. This tank holds reserve water so that the pump is not required to run every time water is needed. Larger pressure tanks extend the life of a pump because they reduce the amount of work a pump must do. When pumps have to frequently cut on and off, they wear out more quickly. Pressure tanks solve this problem.

Another function of a pressure tank is to allow a water system to operate under a higher water pressure. Without a pressure tank, water pressure will be only as good as the pump that is pumping the water. With a pressure tank, the water pressure can be adjusted to meet all needs. Don't accept a well installation that doesn't include a pressure tank of an adequate size.

After the installation of a pump system, you may get frantic calls from customers at all hours of the day and night when they lose their water. You will probably refer these problems to the person who installed the pump for you, but if you know enough about pumps, you might be able to troubleshoot the problem right over the phone with your customer. At the very least, having a working knowledge of pump problems will make it easier for you to understand and deal with problems as they come up. To prepare you for all the common types of complaints you might receive, let's look at some troubleshooting tips for water pumps.

Troubleshooting jet pumps

Because there are many differences between jet pumps and submersible pumps, you will no doubt run into a variety of problems. The following section highlights some of the areas that might need troubleshooting (Table 14-1).

Will not run

A pump that will not run can be suffering from one of many failures. The first thing to check is the fuse or circuit breaker to the circuit. If the fuse is blown, replace it. When the circuit breaker has tripped, reset it. This is something you could ask your customer to check.

When the fuse or circuit breaker is not at fault, check for broken or loose wiring connections. Bad connections account for a lot of pump failures. It is possible the pump won't run due to open contacts on the motor overload protection device. This is usually a temporary condition that corrects itself.

The pump may also not run if it is attempting to operate at the wrong voltage. Test the voltage with a volt-ammeter. The power must be on when this test is conducted. With the leads attached to the meter and the meter set in the proper voltage range, touch the black lead to the white wire and the red lead to the black wire in the disconnect box near the pump. Test both the incoming and outgoing wiring.

Your next step in the testing process should be at the pressure switch. For this test, the black lead should be placed on the black wire and the red lead should be put on the white wire. There should be a plate on the pump that identifies the proper working voltage. Your test should reveal voltage that is within 10% of the recommended rating.

Table 14-1. Troubleshooting table for jet pumps

Symptoms	Probable cause
Won't start	No electrical power Wrong voltage Bad pressure switch Bad electrical connection Bad motor Motor contacts are open Motor shaft is seized
Runs, but produces no water	Need to be primed Foot valve is above the water level in the well Strainer clogged Suction leak
Starts and stops too often	Leak in the piping Bad pressure switch Bad air control valve Waterlogged pressure tank Leak in pressure tank
Low water pressure in pressure tank	Strainer on foot valve is partially blocked Leak in piping Bad air charger Worn impeller hub Lift demand is too much for the pump
Pump does not cut off when working pressure is obtained	Pressure switch is bad Pressure switch needs to be adjusted Blockage in the piping

An additional problem that you may encounter is a pump that is mechanically bound. You can check this by removing the end cap and turning the motor shaft by hand. It should rotate freely.

A bad pressure switch can cause a pump to not run. To check this, remove the cover from the pressure switch and locate the two springs, one tall and one short. These springs are depressed and held in place by individual nuts. The short spring is preset at the factory and should not need to be adjusted. This adjustment controls the cut-out sequence for the pump. If this control does need to be adjusted, you can increase the cut-out pressure by turning the nut down and lower the cut-out pressure by loosening the nut. Typically the long spring is used to change the cut-in and cut-out pressure for the pump. If you want to set a higher cut-in pressure, turn the nut tighter to depress the spring further. To reduce the cut-in pressure, you should loosen the nut to allow more height on the spring. If the pressure switch fails to respond to the adjustments, it should be replaced.

It is also possible that the tubing or fittings on the pressure switch are plugged. Take the tubing and fittings apart and inspect them. Remove any obstructions and reinstall them.

The last possibility for the pump failure is a bad motor. You use an ohmmeter to check the motor, but only when the power to the pump is turned off. Start checking the motor by disconnecting the motor leads, called L1 and L2. The instructions here are for Goulds pumps with motors rated at 230 volts. When you are conducting the test on different types of pumps, refer to the manufacturer's recommendations.

Set the ohmmeter to R×100 and adjust the meter to zero. Put one of the meter's leads on a ground screw. The other lead should systematically be touched to all terminals on the terminal board, switch, capacitor, and protector. If the needle on your ohmmeter doesn't move as these tests are made, the ground check of the motor is okay.

The next test to be conducted is for winding continuity. Set the ohmmeter to R×1 and adjust it to zero. To conduct the test, place a thick piece of paper between the motor switch points and the discharge capacitor.

You should read the resistance between L1 and A to see that it is the same as the resistance between A and yellow. The reading between yellow to red should be the same as L1 to the same red terminal.

The next test is for the contact points of the switch. Set the ohmmeter to R×1 and adjust it to zero. Remove the leads from the switch and attach the meter leads to each side of the switch. You should see a reading of zero. If you flip the governor weight to the run position, the reading on your meter should be infinity.

Now let's check the overload protector. Set your meter to R×1 and adjust it to zero. With the overload leads disconnected, check the resistance between terminals one and two and then between two and three. If you get a reading of more than one, replace the overload protector.

The capacitor can also be tested with an ohmmeter. Set the meter to R×1000 and adjust it to zero. With the leads disconnected from the capacitor, attach the meter leads to each terminal. When you do this, you should see the meter's needle go to the right, then drift slowly to the left. To confirm your reading, switch positions with the meter leads and see if you get the same results. A reading that moves toward zero or a needle that doesn't move at all indicates a bad capacitor.

I realize the instructions I've just given you may seem quite complicated. In a way, they are. Pump work can be very complex. I recommend that you leave major troubleshooting to the person who installed your problem pump. If you are not familiar with controls, electrical meters, and working around electrical wires, you should not attempt many of the procedures I am describing. The depth of knowledge I'm providing may be deeper than you ever expect to use, but it will be here for you if you need it.

Runs but gives no water

When a pump runs but gives no water, you have several possible problems to check out. Let's take a look at each troubleshooting phase in logical order.

The first consideration should be that of the pump's prime. If the pump or the pump's pipes are not completely primed, water will not be delivered. For a shallow-well pump, you should remove the priming plug and fill the pump completely with water. You may want to disconnect the well pipe at the pump and make sure it is holding water. If not, you could spend considerable time pouring water into a priming hole only to find out that the pipe was not holding the water.

For deep-well jet pumps, you must check the pressure-control valves. The setting must match the horsepower and jet assembly used, so refer to the manufacturer's recommendations.

Turning the adjustment screw to the left will reduce pressure, while turning it to the right will increase pressure. When the pressure-control valve is set too high, the air-volume control cannot work. If the pressure setting is too low, the pump may shut itself off.

If the foot valve or the end of the suction pipe has become obstructed or is suspended above the water level, the pump cannot produce water. Sometimes shaking the suction pipe will clear the foot valve and get the pump back into normal operation. If you are working with a two-pipe system, you will have to pull the pipes and do a visual inspection. If the pump you are working on is a one-pipe pump, you can use a vacuum gauge to determine if the suction pipe is blocked.

If you install a vacuum gauge in the shallow-well adapter on the pump, you can take a suction reading. When the pump is running, the gauge will not register any vacuum if the end of the pipe is not below the water level or has a leak.

An extremely high vacuum reading, such as 22 inches or more, indicates that the end of the pipe or the foot valve is blocked or buried in mud. It can also indicate the suction lift exceeds the capabilities of the pump.

A common problem when the pump runs without delivering water is that there is a leak on the suction side of the pump. You can pressurize the system and inspect it for such a leak.

The air-volume control can be at fault when a pump runs dry. If you disconnect the tubing and plug the hole in the pump, you can tell if the air-volume control has a punctured diaphragm. If plugging the pump corrects the problem, you must replace the air-volume control.

Sometimes the jet assembly will become plugged up. When this happens with a shallow-well pump, you can insert a wire through the ½-inch plug in the shallow-well adapter to clear the obstruction. With a deep-well jet pump, you must pull the piping out of the well and clean the jet assembly.

An incorrect nozzle or diffuser combination can result in a pump that runs but produces no water. Check the ratings in the manufacturer's literature to be sure the existing equipment is appropriate.

The foot valve or an in-line check valve could be stuck in the closed position. This type of situation requires a physical inspection and the probable replacement of the faulty part.

Cycles too often

When a pump cycles on and off too often, it can prematurely wear itself out. This happens for a variety of reasons. For example, any leaks in the piping or pressure tank would cause frequent cycling of the pump.

The pressure switch may be responsible for a pump that cuts on and off too often. If the cut-in setting on the pressure gauge is too high, the pump will work harder than it should.

If the pressure tank becomes waterlogged, or filled with too much water and not enough air, the pump will cycle frequently. If the tank is waterlogged, it will have to be recharged with air. This would also lead you to suspect that the air-volume control is defective.

An insufficient vacuum could cause the pump to run too often. If the vacuum does not hold at 3 inches for 15 seconds, it might be the problem.

The last thing to consider is the suction lift. It's possible that the pump is getting too much water and creating a flooded suction. This can be remedied by installing and partially closing a valve in the suction pipe.

Won't develop pressure

Sometimes a pump produces water but does not build the desired pressure in the holding tank. Leaks in the piping or pressure tank can cause this condition to occur. If the jet or the screen on the foot valve is partially obstructed, the same problem may result.

A defective air-volume control may prevent the pump from building suitable pressure. You can test for this by removing the air-volume control and plugging the hole where it was removed. If this solves the problem, you know the air-volume control is bad.

A worn impeller hub or guide vane bore could result in a pump that will not build enough pressure. The proper clearance should be 0.012 on a side or 0.025 diametrically.

With a shallow-well system, the problem could be caused by the suction lift being set too high. You can test for this with a vacuum gauge. The vacuum should not exceed 22 inches at sea level. Deep-well jet pumps require you to check the rating tables to establish their maximum jet depth. Also with deep-well jet pumps, you should check the pressure-control valve to see that it is set properly.

Switch fails

If the pressure switch fails to cut out when the pump has developed sufficient pressure, you should check the settings on the pressure switch. Adjust the nut on the short spring and see if the switch responds. If it doesn't, replace the switch.

Another cause for this type of problem could be debris in the tubing or fittings between the switch and pump. Disconnect the tubing and fittings and inspect them for obstructions.

Troubleshooting submersible pumps

There are some major differences between troubleshooting submersible pumps and jet pumps (Table 14-2). One of the most obvious differences is that jet pumps are installed outside of wells and submersible pumps are installed below the water level in the wells.

There are times when a submersible pump must be pulled out of a well. This can be quite a chore. Even with today's lightweight well pipe, the strength and endurance needed to pull a submersible pump up from a deep well is considerable. Plumbers that regularly work with submersible pumps often have a pump-puller to make removing the pumps easier.

Table 14-2. Troubleshooting submersible pumps

Symptoms	Probable cause
Won't start	No electrical power
	Wrong voltage
	Bad pressure switch
	Bad electrical connection
Starts, but shuts off fast	Circuit breaker or fuse is inadequate
	Wrong voltage
	Bad control box
	Bad electrical connections
	Bad pressure switch
	Pipe blockage
	Pump is seized
	Control box is too hot
Runs, but does not produce water, or only produces a small quantity	Check valve stuck in closed position
	Check valve installed backwards
	Bad electrical wiring
	Wrong voltage
	Pump is sitting above the water in the well
	Leak in the piping
	Bad pump or motor
	Broken pump shaft
	Clogged strainer
	Jammed impeller
Low water pressure in pressure tank	Pressure switch needs to be adjusted
	Bad pump
	Leak in piping
	Wrong voltage
Pump runs too often	Check valve stuck open
	Pressure tank is waterlogged and needs air injected
	Pressure switch needs to be adjusted
	Leak in piping
	Wrong size pressure tank

When a submersible pump is pulled, you must allow for the length of the well pipe when planning the direction to pull from and where the pipe and pump will lay once removed from the well. It is not unusual to have to pull 100 to 200 feet or more of well pipe because of the well's depth.

It is important when pulling a pump or lowering one back in that the electrical wiring does not rub against the well casing. If the insulation on the wiring is cut, the pump will not work properly. Let's look now at some specific troubleshooting situations.

Won't start

Pumps that won't start may be the victim of a blown fuse or tripped circuit breaker. If these conditions check out okay, turn your attention to the voltage. In the following scenarios, Goulds pumps and Q-D-type control boxes are used.

To check the voltage, remove the cover of the control box to break all motor connections. Be advised: Wires L1 and L2 are still connected to electrical power. These are the wires running to the control box from the power source.

Press the red lead from your voltmeter to the white wire and the black lead to the black wire. Keep in mind that any major electrical appliance, such as a clothes dryer, that might be running at the same time as the pump should be turned on while you are conducting your voltage test. The intent is to measure the voltage level under normal loaded conditions.

Once you have a voltage reading, compare it to the manufacturer's recommended ratings. For example, with a Goulds pump that is rated for 115 volts, the measured volts should range from 105 to 125. A pump with a rating of 208 volts should measure a range from 188 to 228 volts. A pump rated at 230 volts should measure between 210 and 250 volts.

If the voltage checks out okay, check the points on the pressure switch. If the switch is defective, replace it.

The third likely cause for the pump not starting could be a loose electrical connection in the control box, the cable, or the motor. Troubleshooting for this condition requires extensive work with your meters.

To begin the electrical troubleshooting, look for electrical shorts by measuring the insulation resistance. You will use an ohmmeter for this test and the power to the wires you are testing should be turned off.

Set the ohmmeter scale to R×100K and adjust it to zero. You will be testing the wires coming out of the well from the pump at the well head. Put one of the ohmmeter's leads to any one of the pump wires and place the other ohmmeter lead on the well casing or a metal pipe. As you test the wires for resistance, you will need to know what the various readings mean, so let's examine this issue.

You will be dealing with normal ohm values and megohm values. Insulation resistance does not vary with ratings. Regardless of the motor, horsepower, voltage, or phase rating, the insulation resistance remains the same.

A new motor that has not been installed should have an ohm value of 20,000,000 or more and a megohm value of 20. A motor that has been used, but is capable of being reinstalled, should produce an ohm reading of 10,000,000 or more and a megohm reading of 10.

Once a motor is installed in the well, which will be the case in most troubleshooting, the readings will be different. A new motor installed with its drop cable should give an ohm reading of 2,000,000 or more and a megohm value of 2. An installed motor in a well that is in good condition will present an ohm reading of between 500,000 and 2,000,000, with a megohm value between 0.5 and 2.

A motor that gives a reading in ohms of between 20,000 and 500,000 and a megohm reading of between 0.02 and 0.5 may have damaged leads or have been hit by lightning, but don't pull the pump yet.

You should pull the pump when the ohm reading ranges from 10,000 to 20,000 and the megohm value drops to between 0.01 and 0.02. These readings indicate that the motor or cables are damaged. A motor in this condition may run, but not for very long.

When a motor has failed completely or the insulation on the cables has been destroyed, the ohm reading will be less than 10,000 and the megohm value will be between 0 and 0.01.

With this phase of the electrical troubleshooting done, we are ready to check the winding resistance. You must refer to charts as a reference for correct resistance values. Adjustments will also have to be made if you are reading the resistance through the drop cables. I'll explain more about this in a moment.

If the ohm value is normal during your test, the motor windings are not grounded and the cable insulation is intact. When the ohm readings are below normal, either the insulation on the cables is damaged or the motor windings are grounded.

To measure winding resistance with the pump still installed in the well, you have to allow for the size and length of the drop cable. Assuming you are working with copper wire, you can use the following figures to obtain the resistance of cable for each 100 feet in length and ohms per pair of leads:

Cable size (AWG)	Resistance
14	0.5150
12	0.3238
10	0.2036
8	0.1281
6	0.08056
4	0.0506
2	0.0318

If aluminum wire is being tested, the readings will be higher. Divide the ohm readings above by 0.61 to determine the actual resistance of aluminum wiring.

If you pull the pump and check the resistance for the motor only, not with the drop cables being tested, you will use different ratings. You should refer to a chart supplied by the manufacturer of the motor for the proper ratings.

When all the ohm readings are normal, the motor windings are fine. If any of the ohm values are below normal, the motor is shorted. An ohm value that is higher than normal indicates that the winding or cable is open or that there is a poor cable joint or connection. Should you encounter some higher than normal ohm values while others are lower than normal, you have found a situation where the motor leads are mixed up and need to be attached in their proper order.

If you want to check an electrical cable or a cable splice, you need to disconnect the cable and have a container of water, such as a bathtub, in which to submerge the cable.

Start by submerging all but the two ends of the cable in the water. Set your ohmmeter to R×100K and adjust it to zero. Put one of the meter leads on a cable wire and the other to a ground. Test each wire in the cable using this procedure.

If at any time the meter's needle goes to zero, remove the splice connection from the water and watch the needle. If the needle falls back to give no reading, there is a leak in the splice.

Once a splice is ruled out, you have to test sections of the cable in a similar manner. In other words, once you have activity on the meter, you should slowly remove sections of the cable until the meter settles back into a no-reading position. When this happens, you have found the section that is defective. At this point, you can either cover the leak with waterproof electrical tape and reinstall the cable or just replace the cable.

Will not run

A pump that will not run can require extensive troubleshooting. Start with the obvious and make sure the fuse is not blown or the circuit breaker is not tripped. Also check to see that the fuse is the proper size.

Incorrect voltage can cause a pump to fail. You can check the voltage as described in the previous electrical troubleshooting section. Loose connections, damaged cable insulation, and bad splices, as discussed, can prevent a pump from running.

The control box can have a lot of influence on whether or not a pump will run. If the wrong control box has been installed or if the box is located in an area where temperatures rise to over 122 degrees F, the pump may not run.

When a pump will not run, you should carefully check the control box. I will be describing the steps for a quick-disconnect-type box. Start by checking the capacitor with an ohmmeter. First, discharge the capacitor before testing. You can do this by putting the metal end of a screwdriver between the capacitor's clips. Set the meter to R×1000 and connect the leads to the black and orange wires coming out of the capacitor case. You should see the needle start toward zero and then swing back to infinity. Should you have to recheck the capacitor, reverse the ohmmeter leads.

The next check involves the relay coil. If the box has a potential relay, three terminals, set your meter on R×1000 and connect the leads to red and yellow wires. The reading should be between 700 and 1800 ohms for 115-volt boxes. A 230-volt box should read between 4500 and 7000 ohms.

If the box has a current relay coil, four terminals, set the meter on R×1 and connect the leads to black wires at terminals one and three. The reading should be less than one ohm.

In order to check the contact points, set your meter on R×1 and connect to the orange and red wires in a three-terminal box. The reading should be zero. For a four-terminal box, set the meter at R×1000 and connect to the orange and red wires. The reading should be near infinity.

Now you are ready to check the overload protector with your ohmmeter. Set the meter at R×1 and connect the leads to the black and blue wires. The reading should be at a maximum of 0.5.

If you are checking the overload protector for a control box designed for 1½ horsepower or more, you will set your meter at R×1 and connect the leads to terminals one and three on each overload protector. The maximum reading should not exceed 0.5 ohm.

A defective pressure switch or an obstruction in the tubing and fittings for the pressure switch could cause the pump not to run.

As a final option, the pump may have to be pulled and checked to see if it is bound. There should be a high amperage reading if this is the case.

Doesn't produce water

When a submersible pump runs, but doesn't produce water, several things could be wrong. The first thing to determine is whether the pump is submerged in water. If you find that it is submerged, you must begin your regular troubleshooting.

Loose connections or incorrectly connected wires in the control box could be at fault. The problem could also be related to the voltage or a leak in the piping system.

A check valve could be stuck in the closed position. If the pump was just installed, the check valve might be installed backwards. Other options include a worn pump or motor, a clogged suction screen or impeller, and a broken pump shaft or coupling. If any of these options are suspected, you will have to pull the pump.

Tank pressure

If you don't have enough tank pressure, check the setting on the pressure switch. If that's okay, check the voltage. Next, check for leaks in the piping system. As a last resort, check the pump for excessive wear.

Frequent cycling

Frequent cycling is often caused by a waterlogged tank, as was described in the section on jet pumps. Of course an improper setting on the pressure switch and leaks in the piping can cause a pump to cut on too often. You might find the problem is being caused by a check valve that is stuck in an open position.

Occasionally the pressure tank is improperly sized, causing problems. The tank should allow a minimum of one minute of running time for each cycle.

With the troubleshooting information I've given you here, you may just be more prepared for a pump problem than some people who consider themselves to be professional pump people. As a plumber, I've seen a lot of people who knew how to install a pump but had no idea how to fix it if something went wrong. This is one reason why I've gone into so much detail in this section. I don't expect you to troubleshoot and fix your own pumps, but you now have enough information to see if what you are being told by your pump person makes sense.

15

Conducting rough-in tests and inspections

Rough-in tests and inspections for plumbing systems are not considered the responsibility of a general contractor, although they are responsible for everything that goes on with a job. If your plumbing contractor fails to call for a rough-in inspection, you may find yourself tearing out work that should not have been done until an approved rough-in inspection was made. This can be expensive, embarrassing, and time-consuming.

Most plumbers are pretty good about getting their inspections, especially if it means they don't get paid until they do. I recommend you take this approach, too, when setting up payment schedules for your subcontractors. Tie your progress payments to the delivery of approved inspection slips. This way you are not paying a subcontractor for work that might not pass inspection.

Even though most plumbers are good with the mechanical aspect of their profession, many are lax in their administrative work. I've seen a number of jobs where plumbers continue on with their work before they have completed the inspections on the previous work. I've seen groundworks covered prematurely with concrete and sewers buried before they should have been. These types of mistakes are almost unforgivable. Most of the blame for such mistakes falls on the plumber, but some blame can usually be aimed at the general contractor.

Almost all new plumbing work requires the issuance of a plumbing permit and periodic inspections by a plumbing inspector. Even the replacement of a water heater requires a permit and inspection. General contractors are normally aware of most inspection requirements, but they don't always make the process easy. Access to a job can be a problem for plumbing inspectors. If the plumbing inspector can't get in to perform an inspection, the plumbing contractor will be charged a reinspection fee and time will be lost. Wasted trips might also put inspectors in bad moods, which is something no contractor wants to do. As a general contractor, providing access to a job is your responsibility. The following are a few common problems that can go wrong during the inspection stage of a plumbing job.

Dealing with inspection problems

Inspection problems can cover a broad spectrum, ranging from leaky pipe joints to locked-out inspectors. General contractors can contribute to some of the problems associated with plumbing inspections, but most of the trouble encountered is the fault of the plumber. Regardless of who is at fault, failed inspections cost contractors time and money. So what can be done to reduce the number of failed rough-in inspections? Following are a few steps that can be taken to lower the number of incidents of failed rough-in inspections.

Being there for the inspector

Access to a job is one of the easiest problems to eliminate when it comes to plumbing inspections. Someone should arrange to meet plumbing inspectors on jobs whenever possible. Ideally, this someone should be a plumber. Many times a minor problem is found during an inspection, and it can be fixed and okayed by the inspector on the spot. When this is the case, the job can be passed rather than rejected. This saves everyone time and keeps the plumbing contractor from being charged a reinspection fee.

I make a point of meeting my inspectors as often as possible so that I can build a good relationship with them. Because of these relationships, there have been many occasions when my presence on a job was enough to avoid a rejection slip. A lot of plumbers don't like to meet inspectors. They see it as a waste of time. I see it as a cost of doing business. Inspectors often have questions about the way in which a plumbing system has been installed. If a plumber is there to answer the questions, an approval is likely. When an inspector has no qualified person to talk with about such questions, a rejection is more likely.

If your plumbing contractor refuses to meet the plumbing inspector on the job, you should consider meeting the inspector yourself. The presence of a general contractor can be nearly as helpful as that of a plumber. The main thing is to have someone meet the inspectors. It not only guarantees access, it makes communication easier and rejections less likely.

Rough-in inspections for new construction are a little different than those encountered with remodeling jobs. Construction sites tend to be much more accessible than existing buildings where remodeling is taking place. I feel it is especially important to meet inspectors on remodeling jobs.

When an inspector goes to a job and can't gain access, it has to be irritating. I'm sure you know how it is to have an appointment for an estimate, then find out that the person you were to meet with forgot about the appointment, causing you to waste your valuable time. This same type of feeling must run through the minds of plumbing inspectors. The last thing you want to do is upset the plumbing inspector, so don't miss the appointment or lock the inspector out.

Some contractors hide keys for inspectors. I've done this myself, but I don't like it. If the key gets lost or the inspector has trouble finding it, the inspection is in jeopardy. The best way to solve access problems is to meet your inspectors on the jobs.

Fixing leaks

Leaks are a prime reason for failed plumbing inspections. This, of course, is not a problem that a general contractor should be expected to fix. There are steps, though, that you can take to lower your risk of failing inspections due to slow leaks.

When plumbers put a test on their rough plumbing, they generally find large leaks right away. Either water is dripping or air is hissing, depending on the type of test. It is the slow leak that evades early detection. Many plumbers test their rough-ins the day before a scheduled inspection and some do the test on the same day as the inspection. Either way, a lot of plumbers bring their gauges up to test pressure and leave the job. There isn't much for them to do while waiting for an inspector, so they go on to another job. This leaves the test gauges unattended. If a slow leak is present, the test gauge will lose pressure. By the time an inspector arrives, the test pressure may be below the required testing level. When this happens, an inspector has no choice but to fail a job.

What can you do to avoid rejections due to slow leaks? If you can't convince your plumber to babysit the gauge while waiting for an inspector, you or a field superintendent, if you have one, can check the gauge. If you find a gauge that is not up to test pressure, you can call your plumbing contractor to correct the problem before an inspector arrives.

The problem of checking the gauges may seem trivial, but it is not. Plumbers who test with air pressure may have a faulty air valve on their test rig. Air can leak past the valve and bring the test pressure down. This type of a leak isn't a part of the plumbing system, but an inspector has no way of knowing what caused a system to lose its air or whether the system was even ever tested. The only real solution is to have someone check test pressures prior to an inspection.

Failing code violations

There are all sorts of code violations that can cause a plumbing job to fail inspection. General contractors are not plumbers, so they cannot be expected to know every aspect of the plumbing code. Neither should they be expected to go around behind their plumbing contractors in an effort to find any code violations that may be present. The typical general contractor would not know all the fittings that can be used for changes in direction from a vertical position to a horizontal one. If the contractor was left to take care of last-minute fitting problems before an inspection, you can bet there would probably be code violations due to the general's lack of knowledge in this area.

While you would have to be very familiar with the plumbing code to catch code violations on your own, there are some simple steps for you to take so that you can avoid rejections due to code violations. If you walk around your job and look at the plumbing installation, you might see some mistakes your plumber made. For example, nailplates are required when a pipe is at risk of being hit by a nail. This is the case when a pipe is close to the edge of a partition wall. Plumbers sometimes forget to install these metal plates. If you notice the mistake early enough, you can have it corrected before an inspector comes and fails the job.

Pipe hangers, or the lack of them, are often the cause of rejected rough-ins. The plumbing code is very specific on the spacing requirements for pipe hangers. For example, schedule-40 plastic pipe that is installed horizontally must be supported at intervals not to exceed 4 feet. Copper tubing that is installed horizontally is required to be supported at maximum intervals of 6 feet. If you learn some of these simple requirements, you can reduce the number of rejected plumbing jobs you have in the works.

Improving your attitude

Bad attitudes often result in rejected plumbing jobs. Throughout the years, I've seen a lot of contractors, both generals and plumbers, argue with inspectors. Believe me, this type of behavior is not going to help you. If an inspector has decided to fail a job, your best defense is concern and cooperation. Showing inspectors that you want your job to be done right is a big help. Agreeing to correct problems, even if you don't agree that your job should be failed because of them, is your best course of action. It can be difficult to smile and agree with an inspector who you believe is dead wrong, but it is almost always the best thing to do. I know this from personal experience. In fact, let me tell you a quick story.

I was very young when I got my master plumber's license. Back in those days, I thought I knew it all when it came to plumbing. Although I knew a lot, I didn't know when to give in to inspectors. I had plumbed the drainage system for an island sink on a job. When the inspector came to look at my work, he made several derogatory comments, such as the piping looked like spaghetti. Being a proud plumber, I took offense to his statements. A verbal battle ensued.

The inspector rejected the job. He did this even after I produced a code book and made my case. The plumbing wasn't sloppy, it was installed per the diagram in the code book. I was furious. In my anger, I went to the inspector's supervisor and pled my case. The supervisor agreed to meet me on the job the next day.

When the next day rolled around, I was loaded for bear. The same inspector who had failed my work showed up on the job. His supervisor was right behind him. We all went into the kitchen area and looked at the plumbing. After just a moment, the supervisor agreed that the work was done properly. He overruled the rejection and gave me an approval slip. I felt great.

After that incident, I noticed that many of the plumbing inspectors who came to my jobs were very businesslike, where before they had been friendly. The next year was a tough one for me. My jobs were being gone over with a fine-tooth comb. You might say that I won the battle, but I lost the war. Arguing with inspectors doesn't pay, even if you are right.

Testing plumbing rough-ins

The tests required for plumbing rough-ins are not complicated, but the rules for them vary with the type of plumbing being installed. Concessions are sometimes made for rough-ins done at remodeling jobs. General contractors are not responsible for testing plumbing systems, but they must be sure that the systems are tested and approved.

Checking water services

Water services are frequently installed without any joints. When this is the case, many plumbing inspectors do not require the pipe to be tested. If any test is required, it is usually at the point where the water service connects to the water source. When a test is required for a water service, air or water may be used. Test pressures vary from code to code, but a test at the maximum working pressure expected to be encountered by the water service is usually sufficient.

Water services may not be required to be tested, but most inspectors want to see the pipe before it is buried. If you mistakenly bury a water service before it has been checked, the inspector may force you to dig it up. Inspectors do have this type of authority.

Inspecting sewers

Sewers, like water services, must be inspected before they are buried. Many plumbers test with air, but some still use water. If water is used, a standpipe with a height of 10 feet must extend above the sewer. When water is put into the sewer for testing, it must reach the top of the 10-foot standpipe. An inspector can stand and watch the static water level for 15 minutes before approving the test. Again, test pressures and times vary from code to code, so check your local code if you have questions about required pressures and the amount of time that a system must hold a test.

Testing groundworks

Groundworks are tested in much the same way as sewers. The drain and vent pipes may be filled with either air or water. If water is used, a 10-foot standpipe must rise above the pipes and be filled to the top with water. Water pipes installed in a groundworks should not contain joints. This being the case, most inspectors do not require a pressure test on the water pipes because they are one continuous length of pipe.

Groundworks must be inspected and approved before they are covered. This means that you should not fill a foundation with dirt, sand, gravel, or concrete until you have an approved inspection slip in hand. Failure to wait for an approved plumbing inspection can result in a costly lesson.

I recall one occasion when a project manager accidentally poured concrete floors in six townhouses before the underground plumbing was inspected. Apparently, the project manager thought the plumbing had been inspected. An inspection had been called for, but the inspector didn't make it to the job. The next day, when the plumbing inspector arrived, concrete was being poured. Arrangements were made to avoid destroying the concrete, but it certainly was within the power of the inspector to require the removal of anything covering the plumbing. Don't put yourself in a position like this.

Inspecting interior rough-ins

Interior rough-ins must be inspected and approved before they are covered up with insulation or drywall. The test procedures for an interior rough-in require that all drains and vents be tested, inspected, and approved before they are concealed. If water is used, it must fill the DWV system to a point where water is present at the vents on top of the roof. When air is used, vent pipes must be capped off to allow the entire system to be filled with air. All water piping must also be tested with either air or water.

In the case of remodeling jobs, inspectors will sometimes be more flexible on test conditions. For example, I've had inspectors who allowed me to fill my pipes to the highest fixture in remodeling jobs without having to test vents. This is done due to the complications of plugging up all outlets in an existing system. However, don't assume that your local inspector will allow this type of testing. There is nothing in the plumbing code that guarantees an easier testing procedure for remodeling work.

Running into delays

Delays in a job can be very costly. If your plumbing installation fails inspection, it can set off a chain reaction. For general contractors, this can mean making several phone calls to reschedule subcontractors. Once a general contractor has to break an established schedule, all sorts of problems can pop up. For example, an insulation contractor might be booked up except for the time scheduled to do your job. If the job is delayed because of a failed plumbing inspection, the insulation crews might not show up for days. This, of course, affects the schedule for the drywall contractor. Being caught in the middle of this domino-effect is no fun; I know because I've been there.

Some jobs will be rejected even when the best efforts are put forth to make sure they pass. Most jobs, however, fail inspections because of someone's negligence. This can and should be avoided. If your plumbing contractor frequently fails inspections, you should consider getting a new plumber.

Getting the slip

When your job passes inspection, make sure you get a copy of the approval slip. You need written proof of inspection approvals for your job files. Being told that a job passed inspection just isn't good enough. If I were you, I wouldn't pay a plumber for a rough-in until an approved inspection slip was produced. Sometimes inspectors don't leave slips on jobs. You can call to confirm that an inspection was done and that approval was granted, but you should request written documentation of this fact. Having approved inspection slips on file provides protection for you, so make sure you get and keep them.

16

Supervising final inspections, callbacks, and warranty work

Final inspections, callbacks, and warranty work are three worrisome aspects of plumbing for general contractors. Final inspections are done at a crucial time and any failed inspection could delay the project. Callbacks and warranty work are done to correct problems with no opportunity to charge for your time. Although both are often considered to be the same, I separate the two into different categories because, in my estimation, warranty work involves problems with a product, such as a faulty faucet, and callbacks are the result of poor work habits, as a rule.

General contractors don't have to arrange final plumbing inspections, but they do have to live with the results. If an inspection is failed, the job is stalled, and customers are likely to be angry. Customers hold general contractors responsible for callbacks and warranty work, even though the problem is more closely related to the plumber who did the work and supplied the materials. The bottom line is this: General contractors are going to get the phone calls and grief from customers.

I have never met a contractor who enjoyed soothing the bad attitudes of dissatisfied customers. It is a part of the profession that must be dealt with from time to time, but I don't think any contractor likes doing it. I know I don't. The best way to please customers is to keep them from getting upset. This means stopping problems before they happen. You can't always do this, but there are many ways to avoid problems.

Passing final inspections

Final inspections represent a glorious time when they are passed, but the joy can turn sour if the inspection is failed. By the time a contractor reaches the end of a job and final inspections, everyone is ready to move on to another job. Customers want to move into their new houses, remodeling customers want their homes to get back on a regular schedule, and commercial customers want to put their building into operation so that they can start making money. When these desires are delayed, customers may become difficult.

There is no viable excuse for failing a final plumbing inspection. If the rough-in work passed inspection, so should the final work, but this is not always the case. Many jobs fail in their final inspections for many reasons, but the most common relates to negligence.

What can cause a final plumbing inspection to fail? Leaks are a common reason for the rejection of final inspections. Poor workmanship, such as toilets that are installed crooked, can be a reason to receive a rejection slip. Forgetting to install vacuum breakers on hose bibbs can also cause an inspector to reject a job. There are, to be sure, a host of reasons for rejection slips.

Can you, as a general contractor, do anything to prevent rejected inspections on final plumbing? Yes, if you are willing to invest a little time in the process. By now, you're probably thinking that it is ridiculous for you to waste your time doing what your plumbing contractor should be doing. I agree with you. The plumber should handle all of these details, but they often don't.

If you assume a posture where you just sit back, let the plumber do the job, and wait to see what happens, you'll probably have more failed inspection slips than approvals. Maybe getting involved is not worth the effort to you, but it is to me because I don't want my customers to lose confidence in me. A failed inspection on the part of one of my subcontractors extends right back to me. Not all general contractors share my position on this issue, but I feel strongly that my customers are depending on me to give them quality work within a scheduled time frame.

If you want to take a positive role in the completion of your jobs, you have to be involved with final inspections. This applies not only to plumbers, but to all of the trades involved. Any final inspection that is failed will create downtime for your customer. To avoid this, you have to put some checks and balances in place.

Conducting your own inspection

One way to reduce rejected plumbing inspections is to conduct your own inspection prior to the official plumbing inspection. This is not as difficult when dealing with finished plumbing as it is with rough-in work. You or one of your field supervisors can check out all the final plumbing installations to see if they are functioning properly. You may not be aware of a few code issues during this final plumbing inspection, but most of this type of inspection has to do with the operation of fixtures.

Testing plumbing fixtures is neither hard work nor complicated, but still many jobs fail because of leaks or malfunctions. How can this be? It's normally a matter of inexperience on the part of the plumber or a lack of interest. Some plumbers simply don't spend the time they should checking out their new installations. When plumbers take this careless attitude, a rejection slip becomes a likely possibility.

Testing fixtures

A final inspection tests the fixtures to see if they are installed with the proper spacing, but if the rough-in work was done properly, fixture spacing shouldn't be a problem. It could be, but it shouldn't be. A quick review of a local code book reveals spacing requirements. At this point, using a tape measure is all that's needed to make sure the fixture spacing is satisfactory.

Workmanship is referenced in the plumbing code. Just because a toilet flushes without leaking and is not too close to another fixture doesn't mean that its installation will not be rejected by a plumbing inspector. For example, if the toilet is not tightly bolted down, an inspector may reject the work. Likewise, when toilets are installed crooked, a rejection slip may be forthcoming. Faucets that are not installed straight and pretty can be reason for rejection. Basic workmanship is judged on appearance. If fixtures look good, workmanship should not be an issue.

Leaks account for most rejections on final plumbing. Sometimes these leaks are pressure leaks, but most of them are drainage leaks. Pressure leaks are easier to find, so fewer of them are left undetected. Drainage leaks can be difficult to spot with a casual glance, and that is one reason why some plumbers stack up rejection slips. You don't have to be a plumber or a plumbing inspector to find leaks. A general contractor like yourself can find the leaks. There are a few tricks of the trade that come in handy when looking for them.

When a final plumbing inspection is conducted, all fixtures should be used to the fullest extent of their capacities. All faucets should be used. Fixtures that hold water should be filled and drained. Pumps should be run and tested. Every aspect of all plumbing fixtures should be checked for defects. This takes a little time, but it is not hard to do.

During my plumbing career, I've seen a lot of leaks. I've also known a lot of plumbers who regularly failed inspections because they didn't completely test their work. Unfortunately, many of these plumbers worked for me. I am not a perfect plumber now, nor have I ever been. My work leaks sometimes, but not very often. Leaks are a part of plumbing, and they have to be expected. But they must also be found and fixed before they cause a job to fail an inspection. Because many plumbers have trouble finding little leaks, let me give you some tips that I've stockpiled over the years. Used properly, my advice can help you reduce your number of failed final plumbing inspections.

Detecting leaks

Leak detection is a troublesome job for some plumbers. There is no reason why they should struggle so hard to find little leaks, but they do. This is, of course, assuming that they are looking for leaks in an aggressive manner. A large numbers of plumbers don't invest much of their time in looking for leaks. They do a quick once-over of the fixtures and call a job finished. It is only when a job fails because of leaks that they get upset. The plumbers can't figure out why they have such bad luck with leaks developing after a test. Well, some leaks do develop over time, but most new installation leaks are there all along; plumbers just don't notice them. If plumbers have trouble finding flaws in their work, you can imagine that someone without a good background in leak detection might become frustrated with their job.

Finding leaks is not normally a difficult task. Some leaks are very apparent, while others are not. Even small leaks are enough to cost you a plumbing inspection, so be sure to fix all of them prior to inspection. Doing this is not always as easy as it sounds. To illustrate how you can find leaks on your jobs, let's look at some common plumbing fixtures and the techniques that I often use to check for leaks.

Checking toilets

Toilets have only three primary locations for leaks. The first location is the supply pipe that delivers water to the toilet. If this pipe is leaking, it is a pressure leak. This means that the leak is under pressure at all times and, therefore, should leak at all times. Leaks are common around compression nuts used on cut-off valves. This is one location where a mystery leak can occur after a thorough inspection. Any movement of the piping that puts stress on the compression fitting can cause a leak. For example, a cleaning crew might come in after your final test-out and hit a compression fitting with a mop. This could be all it would take to make a leak appear. Normally compression fittings don't leak if they are properly put together, but they do need to be checked.

The point where a supply tube enters a toilet is a likely location for a leak. Many times the leak is very slow and won't show up until a puddle of water is found on a floor. A plumber could look for a leak, not see anything, and move on, only to have an inspector find a puddle under the supply tube, resulting in a failed inspection. A thorough inspection of the supply tubes could help you catch these leaks before they cause you problems. For example, when you are checking supply tubes for leaks, you must understand that some leaks are not very visible. Water may be running slowly down the back side of a supply tube. By wiping down all of the tubing and fittings with dry toilet tissue, you will be able to tell if there is a leak. Toilet tissue should be standard equipment in your test kit. It will see a lot of use in the detection of leaks.

After checking all aspects of the water supply, you should check areas where the tank and bowl come together when a two-piece toilet is installed. Toilet tanks are held in place on toilet bowls with either two or three brass bolts. These bolts depend on rubber washers to prevent leaks. If the bolts are not tight enough, water can drip past the bolts at any time while the tank contains water. Wiping the area around these bolts with toilet tissue will make the detection of such leaks easier.

Another place of potential concern is where the tank and bowl meet and water is passed between them. This location will not continuously leak, it will just seep when a toilet is being flushed. This means, of course, that you must flush the toilet in order to discover such a leak. While you are flushing the toilet, watch the edge of the toilet bowl base where it meets the finished floor. If the wax seal between the bowl and its flange is not properly working, water can seep out from under the toilet bowl.

If you check all of the aforementioned areas before the final plumbing inspection, more than likely your toilet will not leak.

Inspecting sinks and lavatories

Sinks and lavatories are both common in residential plumbing, and inspecting them for leaks is pretty simple. Start by turning on the cold water, which should come from the handle on the right. Check to see if this is the case. Sometimes plumbers mix up their water supplies by accident, putting hot on the right and cold on the left. This will cause a job to fail inspection.

After you confirm that cold water is on the right side, turn it off and cut on the hot-water handle. Don't just assume that the left handle will produce hot water; a plumber might have piped cold water to both sides by mistake. Hey, it's happened. Once you get hot water from the left side of the faucet, you know that much of the installation is according to code requirements.

If you want to be especially thorough in your inspection, put a thermometer under the hot water and check the water's temperature. Consult your local code book for temperature limitations. Doing this will eliminate another potential inspection stopper.

Look under the sink or lavatory to inspect the water supplies. Wipe them down with toilet tissue to determine if there are any leaks. Assuming that there are none, you've ruled out another risk of rejection. Now it's time to check the drainage.

When you check the drainage of a sink or lavatory, it's very important that you fill the fixture to its overflow point with water. Some plumbers don't do this, causing them to miss leaks that inspectors find. If you only run water into a sink or lavatory and allow it to drain as it runs, very little pressure will be put on the drainage connection. Filling such a fixture and releasing all of the water at once will produce enough pressure to expose leaks in a drainage system, if they exist.

As water is draining out of a sink or lavatory, inspect all joints in the drainage system beneath the fixture. Once again toilet tissue can reveal leaks that might not be seen with a visual inspection.

If a sink has an accessory fixture, such as a garbage disposal, operate the accessory unit. In the case of a garbage disposal, fill the sink bowl, release the water, and turn on the disposal unit. As the water is draining, check for leaks.

Kitchen sinks are often the location where dishwashers dump their drainage. The dishwashers are normally piped in a manner so that their discharge enters the drainage system under a kitchen sink. A water supply is also frequently found under a kitchen sink that serves a dishwasher. To check these connections, you should remove the access panel on the front of the dishwasher. There will be a water connection behind it. Turn on the dishwasher and have it fill itself with water. Then put the appliance into a draining cycle. Observe all piping for leaks.

If the kitchen contains a refrigerator that is equipped with an ice maker, you should look for the saddle valve that supplies water to the appliance. The valve may be under the kitchen sink or somewhere below the floor in the area of the refrigerator. Saddle valves depend on compression fittings, rubber washers, and bolts to prevent leaks. Very few plumbers bother to check these connections after installing the saddle valve. Leaks often occur here, so check all saddle valves closely.

Testing bathtubs and showers

Bathtubs and showers should be tested to make sure they drain properly and to determine if the hot and cold water is piped to the proper sides. Hot is always on the left and cold on the right. It's surprising how often plumbers cross up their water lines, so don't forget to check each fixture for the location of hot and cold water.

Checking other fixtures

Other fixtures in a house should be checked with the same basic principles given above. As long as you put a fixture through all of its cycles and don't discover leaks or malfunctions, you are well on your way to an inspection approval.

There are a few little things to look for before you conclude your final plumbing inspection. Check all outside hose bibbs to see that they are equipped with backflow protection. Some are sold with this protection built right into them, while others are fitted with a screw-on vacuum breaker. Typically, any device that will accept the threads of a garden hose is required to have a vacuum breaker. This can include washing machine hookups and the spout on a laundry tub. Plumbers sometimes forget to install add-on backflow preventers, which will surely fail a final inspection.

Reducing callbacks

Callbacks are a plumber's nightmare. They disrupt schedules and don't produce any income. In fact, they rob a plumbing company of cash. As a general contractor, there is very little that you can do to prevent plumbing callbacks. If you take an active role in the final inspection of new plumbing installations, you're doing about all you can to prevent callbacks.

Most callbacks are the result of plumbers who are either lazy, negligent, or in a hurry to get to another job. Faulty materials account for some callbacks. Because you are not the direct supervisor of your plumbing crews, you can't stop all possible causes for callbacks.

I see the picture of callbacks from both the general contractor side of the table and as a plumber. Work that I do rarely results in callbacks, but I have had a few. The best plumbers can make a mistake or have something happen that is beyond their control, causing callbacks. Unfortunately, there just isn't much a general contractor can do about unforeseen problems.

I have found only two partial solutions to callbacks from my general contractor's seat. First, closely check over all subcontractor work before paying for it. Second, establish a procedure, in advance, for dealing with callbacks. For instance, set a protocol that your subcontractors must follow when there is a problem with their work. This principle also applies to warranty work.

Doing warranty work

Warranty work is similar to a callback in that nobody makes money off of the work being done. Callbacks are common in the plumbing trade, but warranty work is fairly rare. In my 20 years as a plumber, I've done very little warranty work.

I said early in the chapter that warranty work and callbacks were closely related. They might be considered the same by some people, which is not an entirely incorrect assumption. I separate the two based on the criteria surrounding the work. Going back to fix a leak in a joint that was not properly put together is a callback to me. Returning to a job to replace a toilet handle that has broken off is what I would consider warranty work. Basically I look at it like this, if a plumber is at fault, it's a callback, and when the plumber is not at fault, it's warranty work. This is just my personal way of looking at the issue so that I can keep accurate records for my company.

The odds of a plumbing fixture failing within its warranty period are low. Very few fixtures fail from factory defects. The ones that are defective are usually discovered and substituted during the installation process. It's certainly possible that a new water pump will seize up or that a new faucet will go on the blink, but the frequency of these types of problems is low.

There is, in my opinion, nothing that you can do to prevent warranty work beyond a thorough inspection at the time of installation. The best you can hope to do is to have a quick-response plan in place when problems do surface.

Some subcontractors respond very quickly to warranty work and callbacks. I've always believed this type of work should be put in front of paying work, but not all contractors agree with me. Some subcontractors whom I have dealt with have left me hanging for long periods of time when warranty work was needed. I don't appreciate this type of attitude, and subs who work this way don't work with me for long.

The contracts that I draw up with subcontractors address the issue of callbacks and warranty work. A clause in the contract spells out exactly what is required of the subcontractor in the event emergency or warranty service is needed. I have provisions in my contract that allow me to bring in other subcontractors to correct problems with some other subcontractor's work if the offending subcontractor doesn't respond promptly. Furthermore, I can charge the offending subcontractor for expenses incurred to correct deficiencies. I suggest that you have your attorney draft a similar clause for your contracts.

A fast response to callbacks and warranty work is a general contractor's best defense against disgruntled homeowners. I've found that people are very accepting of accidents, faulty materials, and mistakes as long as corrective action is promptly taken. Experience has also shown me that customers tend to get nasty if they are given a bunch of excuses and promises that are not kept.

You can't eliminate the risk of warranty work, but you can do a lot to ensure that when the time comes for it to be done, you will have a subcontractor available to do it. Establish rules with your subcontractors for dealing with nonpaying work before you have to call on them to perform it. Having an existing agreement on how callbacks and warranty work will be handled is your best course of action.

17

Adding basement baths in existing homes

Adding basement baths in existing homes is a popular type of home improvement that can be quite lucrative. Because of the outstanding income possibilities with this type of remodeling, many contractors aggressively go after this work. If you work with, or plan to work with, the addition of basement bathrooms, there are a few plumbing pointers, highlighted in this chapter, that you should know.

Remodeling contractors often seek work in existing homes. This is a known fact, but did you know that some remodelers target their marketing and advertising to specific types of home improvements? In populous areas where there is enough market demand, some remodelers look for only certain types of work. Kitchen remodeling is often a favorite, followed closely by bathroom remodeling. These two types of remodeling are both very popular and offer the homeowners good returns on their investments.

When I lived in Virginia, my company was known for its expertise in all types of kitchen and bathroom remodeling, although we did other types of work as well. Since I was both a general contractor and a plumbing contractor, there was very good money in kitchen and bathroom remodeling for me. Although my company rarely remodeled basement bathrooms, we frequently installed them.

Basement bathrooms present more challenges to contractors than any other type of bathroom because of their often-unique features. What kind of features you ask? Well, I can't say that I've ever seen an attic bathroom installed where a sump and a sewer-ejector pump was required. Likewise, unless the drainage system for an entire house has to be pumped, an attic bathroom can drain with gravity. This is not always the case with a basement bathroom. I've also never had a rushing stream of water prevent me from installing a bathroom in an attic or addition, but I've had it happen in basements. The homeowner in the house with the underground stream swore that a bulldozer was buried under her neighbor's house. I didn't find any heavy equipment during my installation, but I sure found plenty of water, which didn't make doing the job easy at all. My point is this, basement bathrooms can provide some very unusual obstacles for plumbers and general contractors to overcome.

What do we know so far? We know that although basement bath remodeling projects can be very lucrative, they often provide plenty of surprises and challenges to contractors. Throughout this chapter I will share some of my many basement bathroom remodeling experiences with you. Many of my stories will be very comical, but I'm sure you will also learn information that can open the door to a very profitable venture for you. You will also learn some of the pitfalls associated with basement bathroom remodeling.

Understanding preliminary concerns

There are a number of preliminary concerns that general contractors should think about when estimating the installation of basement bathrooms. It is always a wise idea to have experts accompany you on estimates, especially if a basement bathroom is involved. As stated before, there are a lot of potential pitfalls associated with basement bathrooms, so if you're hoping to profit from their installation, beware.

Ripping up the basement floor

What's under the basement floor? This is one of the first questions you need to have answered before you can give an accurate price for installing a basement bath. Unfortunately, there is no way to know exactly what will be found under a concrete floor. There are, however, some ways to predict the type of plumbing that is present, if there is any. Let me stress the importance of having your plumbing contractor with you when providing estimates for basement bathrooms. There really are a lot of curve balls that can come your way when you take on this type of work.

Assuming that the bathroom you will be installing will contain a new toilet, at least a 3-inch drain must be available. Most codes allow only two toilets to discharge into a 3-inch pipe. This rule may force you into some expensive upgrading of existing pipes, but minimally you must know that there is at least a 3-inch pipe available.

Plumbing is typically not found under basement floors, but it is common for plumbing to exit through the middle of a foundation wall. If this is the case, you will need to install a sump and a sewer-ejector pump in the bathroom. The cost of this equipment alone can add several hundred dollars to the cost of the job.

If you are dealing with an unfinished basement, you should be able to see whether the drain exits through the wall or whether plumbing exists under the concrete floor. Of course you can't see through the floor, but you will know it's under there if you see large-diameter pipes going into the concrete. This is a solid sign that suitable drains may be waiting for you under the floor. If you see floor drains in a basement floor, you may have found evidence of suitable plumbing, but don't count on it. Floor drains can discharge into 2-inch drains, and a basement bath requires a 3-inch drain. You have to locate the main building drain to determine if suitable plumbing exists under a floor for a basement bath.

Outside cleanouts can give you a great indication of what is, or isn't, under a concrete floor. It is not unusual for plumbers to install cleanouts in sewers so that they are close to the foundation of a home, normally within 5 feet.

If you find a cleanout plug outside a basement wall, unscrew the plug and measure the distance to the bottom of the pipe. If this distance is great enough to put the pipe under the footing of the foundation, you know that there is plumbing available under the floor. This doesn't tell you, however, if the drain under the floor is large enough to accept the discharge from additional plumbing. You can assume you are safe if the cleanout riser has a diameter of 4 inches. If the diameter is 3 inches, you may have a problem. Don't just assume that the sewer has a 3-inch diameter if you find a 3-inch riser because it is acceptable practice to install a 3-inch riser on a 4-inch sewer.

You might also get a clue on what size sewer has been used by checking the size of drains in the house. This is pretty easy if you can see exposed pipes. It is not so easy, however, if you can't locate any pipes. If you cannot positively identify the size of the sewer, you had better build a disclaimer into your quote and contract with the customer for the basement bathroom. Failure to put such a disclaimer in your contract can cost you plenty if a 4-inch sewer is not available, assuming that the property has more than one toilet already installed.

When I bid a basement bath and am concerned about the size of the existing sewer, I make sure that my quote is based on the assumption that a 4-inch drain is available within some reasonable location, which I detail in my written quote and contract. By doing this, I avoid the risk of paying an outrageous amount to install a new sewer. The most I stand to lose is the cost of breaking up a floor to determine if a suitable drain is available. This is a risk that I assume, but you could word your disclaimer to recover your costs for this work also. I choose to eat this cost based on past experience, but there is no reason to put yourself at such financial risk.

Finding a suitable drain under the floor

Assuming that a suitable drain is under the floor where you plan to install a basement bath, you are off to a good start, but you're not out of the woods yet. A lot can happen once you begin breaking up a concrete floor. What could go wrong? You can come across more problems than you can imagine when breaking up a concrete floor. The following paragraphs touch on some of these possible problems.

How deep will a suitable drain be buried beneath a floor? It will probably not be buried very deep, but it could be deeper than you expect. This could lead to increased time in making a connection, which could cost you money. If having a pipe a foot or so below a floor is your only problem, you're a lucky contractor.

Will the floor contain rebar and reinforcing wire? It should at least contain wire, and maybe rebar. This will slow your plumbers down a little, but it will not stop them or create such a problem as to cost you a fortune. How thick is the concrete? You should count on it being at least 4 inches thick, but typically it's thicker. Again, this is something that can slow down an installation, but it is still not a huge problem.

Will you encounter underground water beneath the concrete floor? There may or may not be underground water, but if there is it may be your biggest enemy. It has proved to be mine. I told you a little bit about a job where I found a rushing stream under a floor. This story is not fiction. My helper and I broke up a concrete floor and found such a forceful underground stream that the water was cutting a channel through the crushed stone that had been used to fill in the foundation prior to pouring the floor. We used a 2-inch discharge pump to combat the water, but the stream was too much for the pump.

I had put in a typical bid on the job, but it turned out to be anything but normal. With the volume and speed of the water present, it was impossible to use the plastic pipe that I had planned because it was floating away. Furthermore, there was no way to get glue joints to set up in the water. For these reasons, we had to switch gears and use hubless cast-iron pipe for the drains. This was an expensive change in plans. The cost of the cast-iron pipe wasn't the only factor that contributed to my lost profits. Working in fast-moving water is not easy, so the whole job took a lot longer than expected, running up the labor time and cost. Just pouring new concrete was quite a challenge.

The type of circumstances I've just described are fortunately not common. Although this job did cost me most of my profit money, I didn't actually lose any out-of-pocket cash. You have to avoid this type of loss whenever you can.

Hitting rock

Rock can present you with a serious problem if you are planning the installation of a basement bathroom. Like water, solid rock is not always easy to predict. It is possible that you will open up a basement floor and find solid bedrock under the concrete. This either puts an end to your plumbing plans or calls for the use of a jackhammer.

When I began my plumbing career, the first six months were spent using a jackhammer and a shovel to carve channels in bedrock for underground plumbing. This work was being done for new construction, but it proves that concrete floors can be poured just above solid rock.

I've never encountered a remodeling job where I hit solid rock while installing a basement bath, but I know the risk exists. Basement bathrooms that require a pump station could really be a problem if bedrock was encountered. A typical sump for a sewer pump is about 30 inches deep. If your job was sitting on rock, you would have to jackhammer quite a hole to accommodate the sump. Even if a sump pit is not needed, chiseling out paths for pipes in solid rock is going to take some time. There is very little you can do to predict an underlayment of bedrock, but you can put a clause in your contract that will give you some protective options if it is encountered.

Installing the vent

The vent from a basement bathroom with a toilet must have a minimum diameter of 2 inches. This pipe must rise to a point 6 inches above the highest fixture's flood-level rim and tie into a suitable vent or continue upwards until it reaches open air above the roof of the building. A lot of contractors fail to take this work into consideration in bidding a basement bath.

Getting a 2-inch vent up through a finished house is not a simple task. There will probably be some damage to existing walls and a chase might have to be built. Either situation increases the cost of a job. If you fail to add these costs to your quote, you will be out of some profit. It is possible to take the vent up the outside of a home, but this approach is not always acceptable to a homeowner. If the vent is run up the outside of a building, it must be protected from freezing in areas where cold is a factor.

Putting a 2-inch vent up through a one-story house is not a very big deal. The pipe can run up the corner of a closet and be enclosed. When a two-story house is involved, the difficulty level increases. Trying to get the vent lined up with a suitable path can be troublesome. Make sure to remember the need for a vent when you are planning your next basement bathroom.

Adding a bath to a finished basement

Adding a bathroom in a finished basement can create problems for a contractor or plumber. If a basement already has finished walls and ceilings, finding and routing pipes can be very difficult. Water pipes must be located, tapped into, and extended to the new bath. This is simple duty in an unfinished basement, but it can get sticky in a finished one. The same goes for getting access to drains and vents that are above the basement floor. If these pipes are concealed in or behind walls, some cutting and patching will be required. Make sure you are comfortable with the access available to all piping before you commit yourself to a price.

Breaking and patching the concrete floor

Who is going to assume responsibility for breaking and patching the concrete floor? Is this part of the plumber's job or are you on the hook for it? This is an issue you must address with your plumbing contractor. There is no set rule on who does what. As a plumbing contractor, I've done jobs where all the concrete work was done by the general contractor. There have been plenty of times, though, when I took full responsibility for the floor work. This is certainly something you need to work out in advance with your plumber.

Hauling and filling the floor

Who is going to take care of hauling away debris from breaking up the floor and filling the trenches back to grade level? Again, this is something you have to work out with your plumber. I've carried many 5-gallon buckets full of concrete chunks and dirt up out of basements. Hauling out the old concrete and excess dirt is time-consuming, not to mention heavy, work. The contractor responsible for this phase of installation has to be prepared for the job. It takes a strong truck to haul away the rip-out debris produced during the creation of a basement bath.

Filling trenches in after plumbing has been installed is not too difficult. In many cases, existing dirt that was excavated when trenches were dug can be used to fill the substructure. Bagged sand can also be used to fill up trenches. The filling process is much easier than the rip-out process, but it still takes time. You have to determine whether your company or your plumbing contractor is taking on this duty.

Pouring a floor

Pouring a new concrete floor to repair cuts made in an existing one can be fairly simple. I've mixed concrete and sand mix up in a mortar tub many times to patch narrow trenches. I've also had ready-mix concrete trucks back up to a basement and stick their chute through the door to repair a floor.

The worst experience I've ever had with repairing floor cuts came when I was involved with the installation of extensive basement plumbing in an existing home. There were numerous trenches cut in the existing floor, some stretching a long distance. Due to the design of the basement, there was no way to get the chute of a concrete truck into the space that needed repairs. A bucket brigade was formed to haul wet concrete from a ready-mix truck sitting in the driveway of the home to the work area. Six of us toted two 5-gallon buckets of wet concrete into the basement with each trip. It took hours to get all of the concrete into the basement. Fortunately, I was able to stop hauling after a little while because I had to trowel out the concrete and get a finish on it before it dried. Even so, I think my arms grew by at least 2 inches that day from carrying so many buckets of concrete.

Access to a basement can be important in the bidding stage of basement bathrooms. You need access to remove rip-out debris and to haul in concrete. If access is not good, the cost of the job will increase. Keep this in mind as you look at your next estimate for a basement bath.

Realizing a good profit

The money that can be made from installing basement bathrooms can be quite good. I've found that this type of work produces a higher percentage of profit that most other types of home improvements.

Part of the reason for this, I believe, is that the work looks much more difficult than it really is. People pay for work based on perceived value. If a job looks easy, customers don't want to pay a high price to get it done. However, when work looks very difficult, higher prices are expected. Because installing a basement bath looks hard, a high price is usually acceptable.

I should clarify that when I say basement baths are not difficult to install, I'm assuming the installer is experienced and has the proper needed equipment. A strong electric jackhammer will cut through a residential concrete floor with relative ease. Once a floor is opened, the digging for underground pipes is usually not very difficult because most is done in foundation fill. A plumber with an experienced helper should be able to rough-in a basement bath in one day. This timetable applies only to the pipes, not the floor cutting and patching. The fixtures can be set in the trim-out stage in less than half a day. Basically, a plumber can get in and out in two days for a basement bath.

From a general contractor's point of view, a lot more time will be invested. Walls may have to be built and there might be electrical and HVAC work to be done. All of the finish work will take some time. Customers will hear a lot of noise and see a lot of people coming and going. This is what makes the job look expensive. Don't get me wrong, basement bathrooms aren't easy or cheap to install, but they are not as difficult as most consumers think.

It's doubtful that you could choose to specialize in doing nothing but basement bathrooms and make a decent living, but this is one phase of remodeling where the profit margin can be very good. Once you become known as the contractor to call for a basement bath, you might find yourself doing dozens of them each year. As both a plumber and general contractor, I like installing basement baths. Not all contractors share my feelings, but this is a money-making venture that is worth investigating.

18

Plumbing upper-story baths in existing homes

Plumbing upper-story baths in existing homes can be quite a challenge. Getting plumbing pipes up through existing living space with minimal damage requires the expertise of an experienced plumber who is accustomed to doing remodeling work. Plumbers who are not familiar with remodeling work can make a major mess of a house while running their pipes. If you're a remodeling contractor, you need to pick your plumbers carefully.

Homeowners often want to add a bathroom somewhere in their homes. Baths added to a basement or first floor of living space can be troublesome, but they don't present the same potential problems as a bathroom being installed on an upper level. The big challenge when working with upper-story bathrooms is getting drains and water pipes to the proposed bathroom. Drains are the most difficult to install.

Attic conversions often include the installation of a new bathroom. Adding the new bathroom can turn out to be pretty easy, but it can also seem nearly impossible to do without destroying walls and ceilings down below. If a bathroom is being added in existing upstairs space, some destruction is imminent.

How you prepare customers for what will be done during the installation of an upstairs bathroom can have a great effect on how well your job goes. Your prework planning with plumbers will also have a big effect on the success of your jobs. Advance planning is the key to success with upper-story bathrooms.

Working with what's there

Sometimes you can do pretty well with attic conversions by working with what's already there. For instance, if a house is plumbed in compliance with code requirements, there will be at least one 3-inch vent pipe. This pipe will normally pass through an attic on its way out of the roof. You and your plumber might be able to use this pipe as a drain for an attic bath. Don't just assume that this can be done, though. You must first make sure that there is no more than one toilet already discharging into the pipe. If you are lucky enough to find a 4-inch pipe, which is not uncommon in older homes, you might be all set.

If you find a pipe in the attic that might be used as a drain, have your plumber check it out very carefully. Once you have determined that an attic vent can be used, you may still run into problems, such as not being able to cut a vertical vent low enough to allow the installation of a tee fitting, especially if horizontal piping is not possible. Because there is rarely excess headroom in attic conversions, building a raised floor to accommodate underfloor piping is not normally a viable option.

If a vertical vent is coming into an attic through a closet, you can cut out part of the ceiling in the closet to gain lower access to the pipe. You will also need to remove a section of the closet wall, in most cases. Taking this approach allows you to cut and tie into an existing pipe at a point low enough to allow horizontal piping.

It is often easier for a plumber to bring a new pipe to an attic location than it is to fool around with existing pipes. For example, an existing pipe might be present, but at the wrong end of the attic, warranting the addition of new pipes. This is just one reason why the plumber would opt for new pipes, but there are many other reasons. Some general contractors argue with their plumbers on this pipe issue because they feel their plumbers might be trying to pad the job with created work. This could be true, but it probably will not be the case. Plumbers don't usually create extra work for themselves out of preference. If your plumber resists using existing piping, review the reasons for this decision before you get excited.

Creating chases

Creating a chase is usually the easiest way to get a large drain into an upper-level bathroom. The chase might be built into the corner of a closet or room or added to the exterior of a home. Once a chase is built, it can house both drains and water pipes. If you explain to your customers why a chase is needed and offer to make it unobtrusive, they should be willing to accept a chase being built.

Finding water pipe routes

It is much easier to find routes for water pipes than for drains because of their small size and minimal pitch requirements. They are also relatively easy to conceal and can be run horizontally for long distances without any need for substantial grade or pitch. Drains do require a set amount of pitch, which makes them limited in the distance that they may be horizontally run through floor joists.

Vertical water pipes are also much easier to install than vertical drains. If a partition wall doesn't contain fire blocking, it is fairly easy to snake water pipes through them without disturbing their exterior appearance. This can be done with larger pipes, like those used for drains, but the difficulty level increases because of the drain's size and rigidity.

It takes a very good remodeling plumber to drill a hole in both the top and bottom plate of a finished wall so that a pipe can be easily snaked through the holes, but it can be done. I've done it, but it is certainly not easy. Electrical wires, existing plumbing, heating and air-conditioning ducts, and fire blocking can all prevent the successful snaking of pipes.

Installing vertical pipes

Vertical pipes present fewer problems than horizontal pipes, in most cases. While a chase may be needed to get vertical pipes into an upper-story location, the disturbance of existing space is limited. Once a suitable location is determined, installing vertical pipes doesn't take much time. The process should not be particularly disruptive or difficult because the only requirement is a location where the pipe can be installed vertically with its diameter concealed.

Positioning horizontal pipes

Horizontal pipes are more of a problem than vertical pipes because they often run through floor joists, which sometimes have limited access. Floor joist concerns are not a problem in an attic conversion because the pipes can be installed before subflooring is secured to the joist. Water pipes don't require grading, so they can run horizontally for an unlimited length.

Drains are required to have a downward pitch. This, combined with their large diameter, can create some real problems. Holes can be drilled in floor joists, but a certain amount of the top and bottom of a joist must be preserved to maintain structural integrity. While this isn't a problem with water pipes, it can become a problem with drains.

Trying to install a large drain, such as a 3-inch pipe, in a horizontal run through floor joists can prove to be very difficult. This is especially true if the length of the run is long or if the joists being penetrated are small. If we say that holes may not be drilled closer than 1½ inches to either the top or bottom of a floor joist, 3 inches of the joist is unusable. Now, if we say that the joists are made of 2-×-8-inch lumber, which has a vertical measurement of 7½ inches, we have only 4½ inches with which to work. Subtract the 3-inch pipe, which measures 3 inches on the inside so it's more than 3 inches around, and you're left with less than 1½ inches of space. If a standard ¼-inch-per-foot grade is used, this means that the 3-inch pipe can run horizontally for approximately 6 feet.

Assuming that we run a 3-inch pipe as described, the center of the pipe will be about 3 inches below the top of the floor joist when it ends its run. If we try to put a 3-inch elbow on this pipe to accept a closet flange for the toilet installation, the flange will probably be too high above the floor level. The use of a street flange may allow a suitable installation, but height can still be a problem. Now the length of the run is limited and the height requirements pose a problem. This might require the use of larger floor joists or the raising of the bathroom floor.

Are you starting to see some of the reasons why advance planning is needed with an upper-story bath? If you allow your carpenters to frame the floor of an attic conversion with 2-×-8-inch joists before you know where your pipes need to be, you may find that the plumbing can't be installed without major modifications. Because main drains normally need to be brought up in close proximity to the fixtures being served, it is necessary to pick paths for pipes before framing is done.

Setting fixtures

Once you find a way to get pipes to the site of a new bathroom, the general rough-in work and setting fixtures aren't any more difficult than you would expect for a regular bathroom. The key to success with upper-story bathrooms is advance planning. I know you've heard this before, but it is true. If you lay out your plumbing plans in advance, you can head off problems while they are still on paper. This is much easier and less disruptive than when problems pop up during the course of field work.

Talk with your plumbers about how pipes will have to be installed before you bid a job. You may find that there will be a need for larger floor joists or a raised platform. This will, of course, affect the cost of your labor and materials. Once you have a clear picture of how pipes will be routed, you can plan your other work around the plumbing. This type of approach will make your upper-level bathrooms easier to install.

19

Relocating fixtures in existing homes

Relocating fixtures in existing homes can be very simple, but it can also get a bit more complicated. One of the first things to consider about the relocation of existing fixtures is the need for a plumbing permit. With simple relocation, a plumbing permit and official inspection is generally required; however, for normal fixture replacement, no permit or official inspection is necessary. This shouldn't be considered a big deal, but it is something that you should know.

I should note that a permit and official inspection by a code enforcement officer is required for the replacement and relocation of a water heater because of the danger involved if it is improperly installed. This is unusual in that most fixtures are exempt from permits and inspections if a simple replacement is being done. It is very important that you be aware of the rules governing the replacement or relocation of water heaters.

Moving water heaters

Because water heaters are pressurized vessels, there is a possibility that they could explode, causing serious damage to both property and person if moved and installed incorrectly. For this reason, all major plumbing codes require a permit and inspection process for their replacement or relocation. Unfortunately, some plumbers don't play by the rules in this area. I've known a lot of plumbers who have avoided the hassle and expense of getting permits and inspections in association with the replacement or relocation of these units. This is a big risk for the plumber, as well as the general contractor to some extent.

If a water heater is relocated or replaced without using proper procedures, a very dangerous situation can exist. For example, if a pressure-and-temperature relief valve of the proper rating is not installed, a water heater can become a bomb of sorts. Failure to pipe a discharge tube from a relief valve can result in scalding injuries to people. Improper electrical wiring can be responsible for a fire. The installation of a cut-off valve on the hot-water side of a water heater could allow excessive pressure to build up and create an explosion. Obviously, none of these scenarios are good, and if you are involved in a water-heater installation that goes bad, it could mean a lawsuit.

When plumbers decide to cheat the system by avoiding permits and inspections on water heaters, they are taking a very big risk. Most plumbers who cheat don't think about what could happen to them. As a general contractor, you may not even be aware that a permit and inspection is needed. Although this may be understandable, it may not be excusable in a court of law. As a general contractor, you are in a prime position to get sued if a plumbing installation done under your direction goes bad. Because there could be a loss of life, a crippling injury, or extensive property damage associated with a bad water-heater installation, you have to protect yourself. The best way to do this is to insist on having your plumber follow code regulations to the letter. This means getting a permit and inspection for every water heater replaced, relocated, or installed. I'm not a lawyer, so I can't give you legal advice, but I can tell you what my personal procedures are in terms of water-heater work.

In my case, water heaters are installed by my own plumbing company. This makes it easy for me. I insist that permits are obtained for all water-heater work where required. Once a permit is secured, inspections are sure to follow. This, in my opinion, protects me. There is still a risk that a relief valve could fail and a water heater explode, but if this happens, I can produce documentation that the work done by my company was in compliance with code regulations. This doesn't mean that a lawsuit won't be aimed at me, but it does give me a better position. An approved inspection by the local plumbing inspector proves that the installation was properly made. Without an approved inspection slip, I'm sure to be found a negligent plumber who violated the plumbing code.

While I might get sued for a water heater that blows up, my insurance company should pay any founded claims as long as I did the work professionally, properly, and in compliance with the local plumbing code. This might not be the case if I cut corners on the code requirements. I can't afford to gamble on getting sued and losing a major lawsuit, and I doubt if you can either. Aside from the money issue, I would have a difficult time living with myself knowing that my negligence caused damage or injury to a person or property. Permits for water-heater work are not expensive, and required inspections don't require a lot of time. From my perspective, there is no excuse for avoiding the permit process, but there are many good reasons for complying with code regulations.

Relocating toilets

Replacing a toilet does not normally require a permit; however, the relocation of any plumbing fixture typically does. Moving a toilet can be a simple process, but due to the size of the drain pipe needed for a toilet, complications can arise. If a toilet drain will be moved to a joist bay beyond the original location, drilling will be needed. Depending on the height of an existing pipe, drilling floor joists for a horizontal installation of pipe can be a problem. This is similar to the problem discussed in Chapter 18 pertaining to running pipes horizontally through joists.

Cutting existing pipes that serve as toilet drains can pose problems for plumbers who are not in possession of the proper equipment. For example, cutting a cast-iron pipe that is installed between two joists is very difficult without a pair of ratchet cutters. The plumber may have a pair of snap cutters, but they require too much space to operate. Hacksaws will not do when cutting cast-iron pipe because that method is extremely slow, as are reciprocating saws with metal-cutting blades. Having ratchet cutters for this type of work is almost a necessity for a prompt, professional job.

If piping is accessible from below, such as in a crawl space or basement, making a relocation for a toilet is not too difficult. Moving a toilet drain that is embedded in concrete is the most difficult type of job in terms of relocation. As a general contractor, you must assess these types of conditions when you are bidding work. Failure to do so can result in a lot of lost money.

Draining a new sink or lavatory location

Sinks and lavatories are often pretty easy to move because of the small size of their drains. This, however, doesn't mean that there could not be trouble in your path. For example, in your relocation efforts, you could drill through heat ducts or a gang of jack studs under a window. There can be any number of obstacles in the path you hope to use for draining a new sink or lavatory location.

As with toilets, sink and lavatory drains are easy to move when you have adequate access and room to work. Moving pipes horizontally, however, can be a big problem. There are many creative ways to hide drains once they are moved. In the case of a kitchen sink, you might be able to run a drain on the outside of a wall and conceal it with cabinets.

Galvanized pipe was commonly used for small drains and vents at one time, as was cast-iron for large drains and vents. Although there are not many problems with the cast-iron drains and vent, if your plumber will be relocating a drain that is made of galvanized steel pipe, insist on having as much of it as possible replaced with plastic. Because the steel rusts and develops rough spots, stoppages generally occur when the rough spots catch and trap objects that are going down the drain. It would behoove you to make the replacement when all it will cost is the price of the pipe and a little extra labor.

Moving bathing units

Moving bathing units, such as bathtubs and showers, can require some ceiling destruction. If the drains for these fixtures are accessible in a crawl space or basement, there won't be much of a problem, but when the units have finished living space below them, the ceiling is going to have to be opened up to allow for relocation. The small diameters of the drains should allow adequate freedom of movement, but your plumber will have to gain access to them. Depending upon how the initial installation was done, your plumber may have to take out a substantial portion of a ceiling to reach a point low enough to allow the drain to be moved as needed.

It is not uncommon for a plumber to roll fittings upward during the installation of a tub or shower drain because the traps for these fixtures have to be held high in joist bays. By getting the drains high into the joists, there is not a problem with traps being below a ceiling line. Unfortunately, having drains up high can make relocation more of a problem. This is why a plumber may have to work backwards for some ways to find a low point in the drainage that is suitable for relocation.

Spacing the fixtures

Fixture spacing is regulated by major plumbing codes. The spacing is not normally a problem, but it is something that you have to keep in mind. If a customer asks you to move a toilet or add a larger bathtub, you have to be aware that your actions are limited by the confines of your local plumbing code. Purchasing a code book for the plumbing trade can help you with this situation. Fixture spacing is an issue that is easy to understand in the text of all major plumbing codes. You can refer to the book to see if what your customers want is feasible. If you are not inclined to use a code book, you can ask your plumbing contractor.

Converting old pipes

Old pipes can give plumbers trouble during the relocation of fixtures. Most drain and vent pipes can be easily converted to more modern pipes thanks to the use of rubber couplings. When you get into water pipes, however, rubber couplings won't cut it.

Some types of old water pipes can require a plumber to tear out a lot more than what is expected. Threaded pipe, such as galvanized steel or brass, can corrode. If the threads go bad, a plumber must remove pipe until a suitable fitting is found to tie into. Once female threads are found, adapters can be used to convert to any type of pipe. However, getting back to a female thread pattern can require extensive work. In theory, a plumber might have to remove 20 feet of pipe, plus or minus, to find a female thread pattern. This would be unusual in a closed wall, but it is possible. In any event, you would have to destroy more wall or ceiling than previously planned.

If you will be relocating fixtures that are suspected of being served by threaded pipe, you might want to put a disclaimer in your contract to deal with unexpected problems. Not doing this could result in a lot of lost profit, not to mention some very upset customers.

20

Plumbing new room additions

Plumbing new room additions is very similar to plumbing any new construction job; however, you typically connect new water supplies and drains to the existing plumbing system. Tying into old plumbing can present a plumber with problems.

Although most residential plumbing systems can support the addition of new plumbing, it should not be assumed that they can. Someone, typically the plumber, must make sure that existing systems can take on additional responsibilities. As a general contractor, your responsibility is to seek a definitive answer to any question of whether or not existing plumbing can be used to accept additional loads.

Determining the number of toilets

As we have discussed before, major plumbing codes limit the number of toilets that may discharge into a 3-inch pipe to two. This is a simple rule to understand. If a house has a 3-inch sewer, no more than two toilets may be drained into it. There is no mystery about this rule, but it can get you into a lot of trouble.

Some houses do have 3-inch sewers and two existing toilets. When this is the case, a third toilet cannot be added without either running a new sewer for the additional toilet or enlarging the existing sewer. Either of these acts will be expensive. If you fail to see that the installation of a new toilet will require such work, you may have to pay for the work out of your own pocket. Make very sure that any existing plumbing you plan to use can accommodate the changes you are planning to make before you agree to make the changes.

Checking the adequacy of water services

It is possible, but unlikely, for an existing water service to be too small to handle the additional load of new plumbing. Just as with a sewer, you must make sure that the size of an existing water service is adequate to meet the needs of additional plumbing. Although I've never found a water service in a residential application that was too small for proposed additions to the plumbing system, it doesn't mean it can't happen.

Unlike the two-toilet-for-3-inch-sewers rule, determining the size of a water service is more difficult. The issue is addressed in a plumbing code book, but the sizing requirements can be confusing. General contractors can probably use the information provided in a code book to determine if a water service is large enough, but a plumber should be consulted.

The size of a water service is determined by water pressure, the length of the pipe, and the fixture-unit load on the pipe. If an average house has a one-inch water service, you can feel sure that the pipe is large enough to handle more plumbing. Even a ¾-inch water service should be large enough to allow the feeding of extra fixtures, but there might be circumstances where a larger pipe is needed. You might have to enlarge the pipe if you were adding on to a house that was very large, had an extremely long run of water-service piping, or possibly a very low water pressure. If you run across a ½-inch water service, you had better plan on enlarging it or running a new water service for the new plumbing.

Tying into an existing water-distribution system

Assuming that an existing water service is large enough, it can be tapped into for additional water distribution. The actual tap will normally be made on some part of an existing water-distribution system rather than directly onto the water service. When tying into an existing water-distribution pipe, you must again make sure that the pipe being connected to is large enough.

A ¾-inch water distribution pipe should be large enough to allow the feeding of additional fixtures. Unless there is a tremendous amount of fixture units already on the pipe, there should be no problem adding a new bathroom or two. Confirming the ability of a water-distribution pipe to handle new loads will have to be done using sizing criteria found in the plumbing code. This process can get complicated, so it might be in your best interest to consult your plumbing contractor for sizing requirements.

A ½-inch water pipe is restricted to serving no more than two fixtures. For example, you might be able to tap into a ½-inch pipe if you are adding only one new plumbing fixture, but you could not use a ½-inch pipe to supply cold water to a full bathroom. If you are adding more than one new plumbing fixture, you should tie into a ¾-inch or larger pipe.

I have remodeled many old homes in which all of the water-distribution pipes were ½ inch in diameter. By current code requirements, this size pipe is not acceptable. To add new plumbing to such a system might require some upgrading of the old piping. It may be possible to tie into the existing pipes, even though they are too small, but before you attempt this, talk to both your customer and the local plumbing inspector.

A plumbing inspector may consider an old, undersized system to be grandfathered under the code. This simply means that the existing system may be used in its present condition until it fails, but once it does, it must be updated to modern standards. If you are adding to an old system, the fact that the existing system is not in compliance with the code may or may not affect you. Some officials may allow the existing system to remain as is, but prohibit you from tying new pipes into it without bringing some portion of the system up to current code requirements. Other officials might allow the old system to be tapped into without major modifications. This is a question you must have answered by your local plumbing inspector.

Let's assume that a plumbing inspector gives you permission to tap into an undersized system to feed the new plumbing you plan to install. It's best to get written authorization to do this from the code officer, but even then it may not be acceptable to the customer to go ahead with the plans. Tying into an undersized plumbing system and extending new pipes to a new bathroom might result in low water volume or pressure. Your customer may not want this. Even though you have permission from a code officer to proceed in this manner, the customer may feel that you should have followed current code requirements. This situation could get nasty. Before you make any part of your new installation in a way that is not conforming to current code regulations, get written permission from both the local code officer and your customer. It's important that you get authorization in writing. If you wind up in court later, the documents you possess will be your best defense.

Once suitable pipes have been located for a tap-in, your plumber will need room to work. This is rarely a problem with water pipes because very little room is needed to cut or make joints in them. One exception to this, however, could be if the existing pipe is threaded. If this is the case, your plumber will need threads or a female-threaded outlet to work with in order to join the pipes. Not many homes contain threaded water pipes anymore, but it is possible that you could find one that does.

Tapping into existing drains

Tapping into existing drains is a little more difficult than tying into water pipes. If the house you are working with has plastic pipe used as a drain, there shouldn't be a problem. However, if you have to tie into cast-iron pipe, you might run into some trouble. A lot of houses still have cast-iron drains. Cutting and tying into cast-iron pipe requires some working room. If the section of pipe that will be tied into is extremely close to a foundation wall, it might not be possible to get the cutter chain of a ratchet cutter around the pipe. If the pipe has to be cut with a metal-cutting saw blade, extra time and effort will be required. It can be done, but your plumber won't be happy. When you are evaluating old plumbing for tie-in purposes, take mental notes of the access available to and around the pipe.

Getting new plumbing to existing plumbing

When building new room additions, it is sometimes very difficult, if not impossible, to get new drains and water pipes from the addition to existing plumbing in the house. A room addition that is built at some distance from existing drains may be too far away to allow the needed grade on drains that are trying to reach existing plumbing. Design factors may also prohibit tying water pipe and drains into existing plumbing systems. There is almost always some way to make new plumbing work, but at times the options are expensive. Before you quote a room addition that will contain plumbing, you should make certain of how your plumbing will be routed.

Using crawl space foundations

Crawl space foundations provide an opportunity for getting pipes from an addition to the main plumbing system in a house. If the addition will be built with a foundation of equal height to that of the existing foundation, running water pipes under the addition to existing plumbing should not be a problem. The main drain may be another story. Because the drain must maintain a constant, downward grade as it runs from the addition to a point of connection with a building drain or sewer, distance could be your enemy.

It is easy enough to predict if you will have problems maintaining suitable grade on a drain. Measure the distance that the drain must cover. Let's say it's 20 feet. If your plumber will be using a standard grade of ¼ inch per foot, the pipe will have to drop 5 inches during its run. It is possible to use less grade on a pipe under certain circumstances, so you might be able to get by with less drop. Once you know how much the pipe will have to fall, you can measure the difference in height between your connection point and your starting point. It is best to measure from the center of the drains. If the subfloor for the addition and the main house are at the same level, you can measure down from it to the pipe. Assuming that you have 5 or 6 inches, you can make the tie-in. If you have only 3 inches of difference, a new plan of attack must be formulated. This might call for tying into the sewer outside of the home.

Using basements

Houses with basements offer similar opportunities to those discussed with crawl space foundations. As long as your plumber can penetrate the basement wall and gain access to the addition, a tie-in can usually be made. Once again, you must confirm that the drain in the basement is low enough to allow the new drain to be tied in while maintaining sufficient grade.

Dealing with slab foundations

Accessing the plumbing in houses on slab foundations is not easy. If you are building a room addition on a house that sits on a slab, you should consider tying the drain into a sewer out in the lawn. Another alternative might be to run the pipes through the attic or interior walls, but the drain will probably have to be installed in an exterior trench.

Having foundations of varying heights

It is not uncommon for room additions and existing homes to have foundations of varying heights. For example, a house might be sitting on a crawl space foundation while the room addition is built on a slab foundation. If this is the case, you will probably have to make your sewer tie-in outside the foundation walls. Before any plan of action can be put into play, you must assess the location of existing plumbing with respect to the new plumbing.

Adding an independent sewer

An independent sewer is sometimes needed when plumbing a room addition. You've just seen some of the types of conditions that might require installing an individual sewer for just the addition. Sometimes an independent sewer is the easiest way to tie in the plumbing from an addition. Even if it is possible to connect to an existing drainage system, it could very well be easier to run the building drain for an addition to a sewer connection. This would not normally be the case if a road-cut is required, but if you can connect to an existing sewer in the lawn, it might be your best option. The downside to this is, of course, the excavation work. You and your plumber will have to weigh all of your options to arrive at the most cost-effective way to handle your drains.

If existing plumbing is accessible and suitable, the need for an independent sewer will not normally be an issue. When an independent sewer is needed or wanted, you have to take the normal precautions for the digging and filling of the sewer trench. When you can tie into an existing building sewer, one that is near the house and in the lawn, you should not have to worry about paying tap fees. If you plan to tap into a municipal sewer, however, a tap fee will normally be required, even if there was a tap fee paid when the existing sewer was connected to the main sewer. Tap fees can be expensive, so don't overlook this cost if you suspect it is a possibility.

Having no way to drain a bathroom

What would you do if you had sold a room addition and then found out that there was no way to drain the bathroom in the addition? If this were to happen, you would certainly have a mess on your hands and probably a very disgruntled customer. Fortunately, it is very unlikely. If it were to happen, a solution would be to install a pump station. This would obviously add to the cost of a job, but it would be better than having no way to drain a bathroom.

There are only two reasons that I can think of that might create a situation where there would not be some way to drain a bathroom. If an independent sewer had to be used for some reason, you could run into a roadblock if a municipality refused to allow you to tap into a main sewer. Could this happen? I suppose it could, but I've never heard of it. As long as there is an existing sewer serving a home, you should be able to tie into it by some means. Again, a pump might be needed, but it would work.

Houses served by septic systems could present you with a problem that would not be at all easy to fix. For example, if an existing septic system was at full capacity, you could not add a bathroom without expanding the septic system. If you decided to expand the septic system, you'd have to make sure the house lot was large enough to allow for the expansion. A soil test should also be conducted to see if the ground could accept additional plumbing. Running into situations like these could bring your plumbing plans to a dead stop. If you will be working with a house that has a private sewage disposal system, be very careful.

In a worst-case scenario, a concrete holding tank might be allowed for use in place of a standard septic system. If you are unable to use any normal means of sewage disposal, a holding tank might be the answer. Having to pump the tank out each time it is full may be a nuisance and expensive, but it might be the only viable option.

Ensuring available connections

Available connections are the only unique risk involved with plumbing new additions. As long as you go into these types of jobs with your eyes open, you shouldn't encounter many problems. It is a very good idea to take a plumbing contractor with you to look over existing conditions that might affect your work. Once you make a contract commitment, you can be held to it, so don't make contractual promises you can't keep. Make sure that you have access and adequate provisions for the proposed plumbing in your next job.

A

Plumbing code

Take some time to look over the information in Figs. A-1 through A-15 and in Tables A-1 through A-48. It can help you learn much about the plumbing code. You will also note some of the differences between the various codes. Reviewing these tables makes the rest of this book more useful. Because there are three major plumbing codes, I've laid out many of the following tables with zone numbers. Refer to Tables A-1, A-2, and A-3 to see which tables apply to your region.

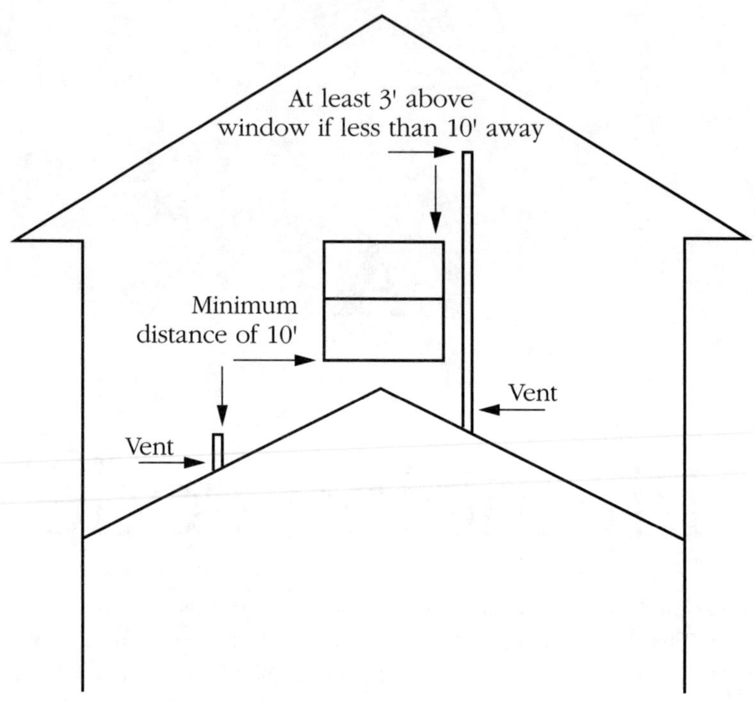

A-1 *Guidelines for vent termination*

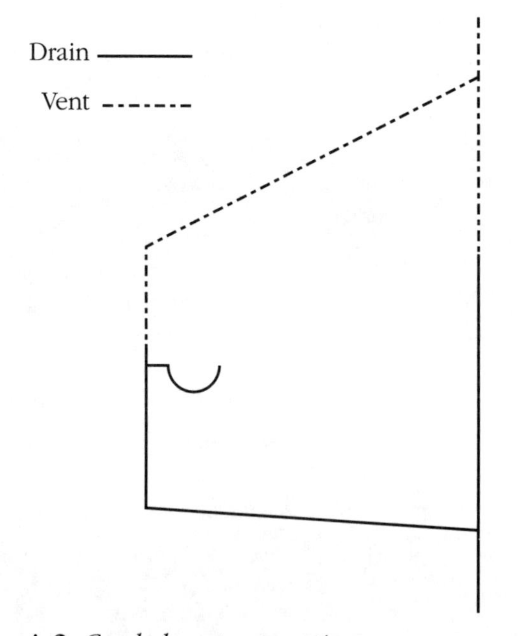

A-2 *Graded vent connection*

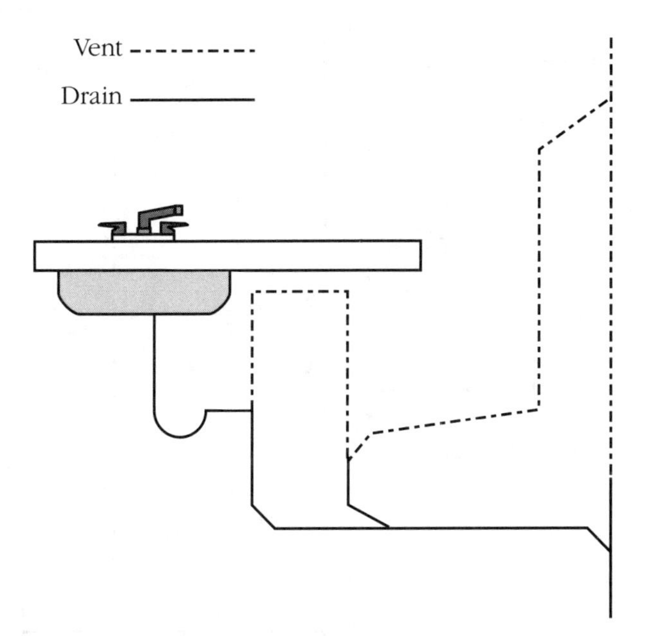

A-3 *Island vent*

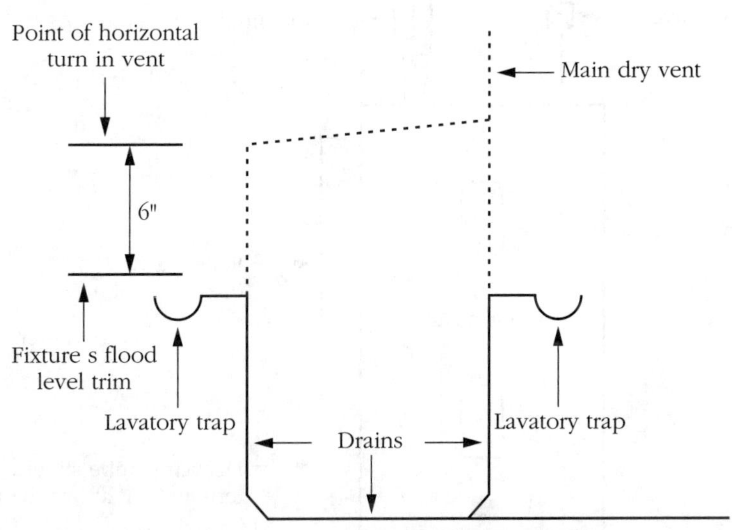

A-4 *Height requirement for turning a vent to a horizontal run*

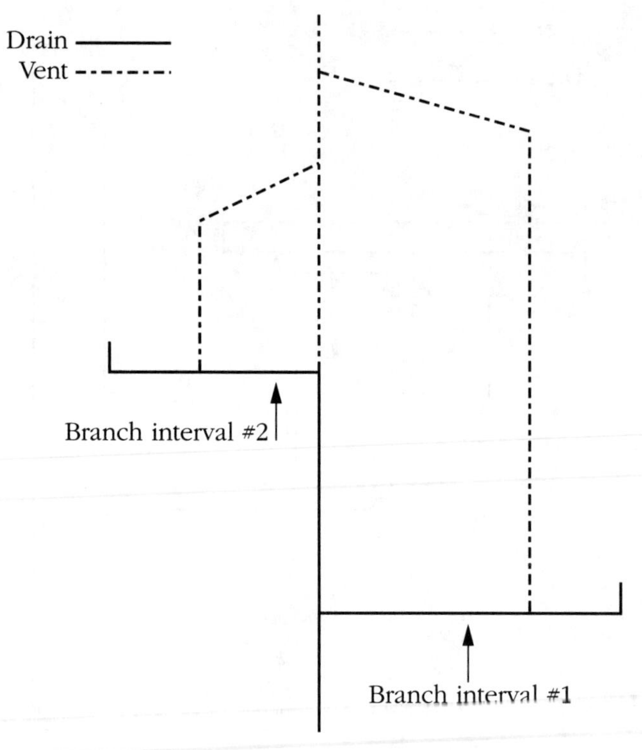

A-5 *Drainage stack with two branch intervals*

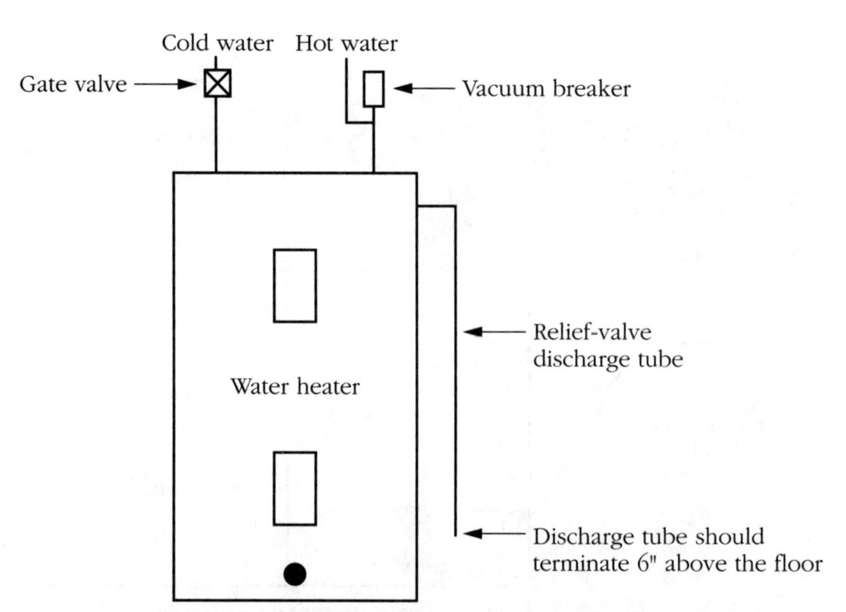

A-6 *Typical water heater layout*

A-7
A long-turn fitting, which can be used to turn horizontal drains on a right angle

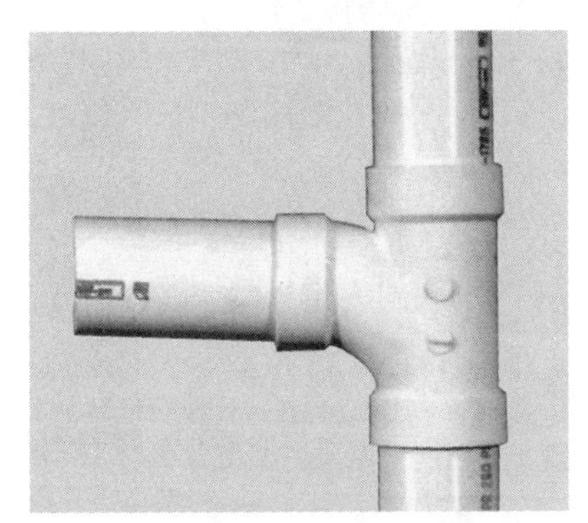

A-8
A short-turn fitting, which should not be used to turn a horizontal drain on a right angle

A-9 *A vent pipe that should be protected with a nailplate*

A-10 *A nailplate being used to protect two water pipes*

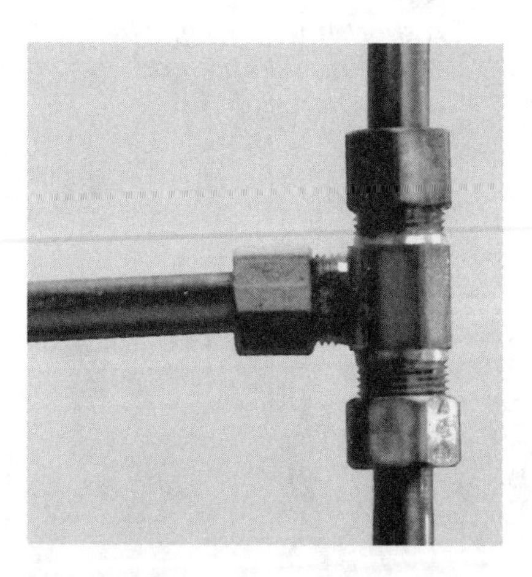

A-11
A compression tee fitting

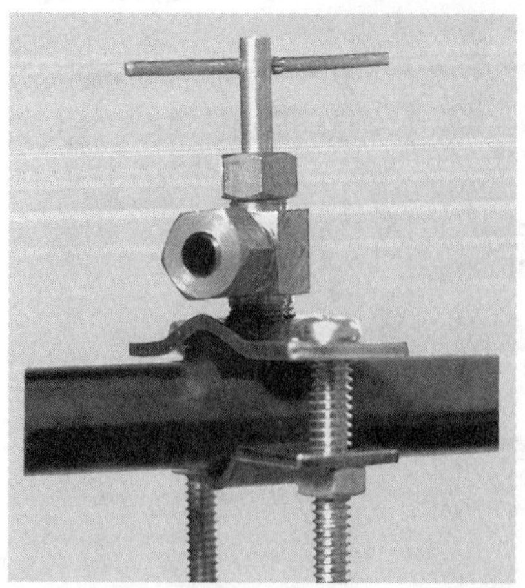

A-12
A saddle valve that might be used to supply water to an ice maker

A-13
A backflow preventer

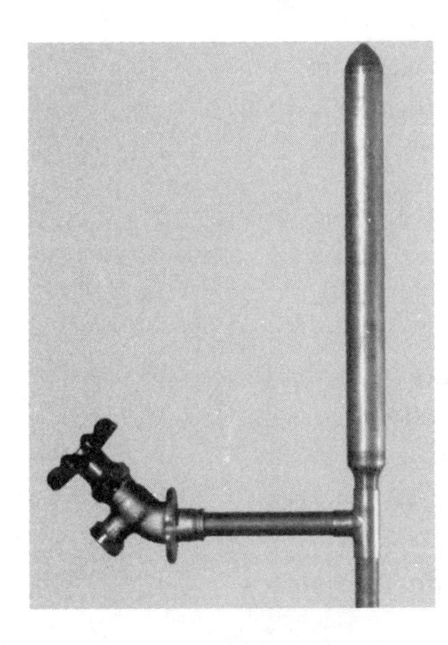

A-14
An air chamber installed above a fixture to reduce the banging of pipes, known as a water hammer

A-15
A temperature and pressure-relief valve which might be installed on a water heater

Table A-1. Zone-One states

Washington
Oregon
California
Nevada
Idaho
Montana
Wyoming
North Dakota
South Dakota
Minnesota
Iowa
Nebraska
Kansas
Utah
Arizona
Colorado
New Mexico
Indiana
Parts of Texas

Table A-2. Zone-Two states

Alabama
Arkansas
Louisiana
Tennessee
North Carolina
Mississippi
Georgia
Florida
South Carolina
Parts of Texas
Parts of Maryland
Parts of Delaware
Parts of Oklahoma
Parts of West Virginia

Table A-3. Zone-Three states

Virginia
Kentucky
Missouri
Illinois
Michigan
Ohio
Pennsylvania
New York
Connecticut
Massachusetts
Vermont
New Hampshire
Rhode Island
New Jersey
Parts of Delaware
Parts of West Virginia
Parts of Maine
Parts of Maryland
Parts of Oklahoma

Table A-4. Vent sizing for Zone Three

Drain pipe size	Fixture-unit load on drain pipe	Vent pipe size	Maximum developed length of vent pipe
1½"	8	1¼"	50'
1½"	8	1½"	150'
1½"	10	1¼"	30'
1½"	10	1½"	100'
2"	12	1½"	75'
2"	12	2"	200'
2"	20	1½"	50'
2"	20	2"	150'
3"	10	1½"	42'
3"	10	2"	150'
3"	10	3"	1,040'
3"	21	1½"	32'
3"	21	2"	110'
3"	21	3"	810'

Table A-4. Continued

Drain pipe size	Fixture-unit load on drain pipe	Vent pipe size	Maximum developed length of vent pipe
3"	102	1½"	25'
3"	102	2"	86'
3"	102	3"	620'
4"	43	2"	35'
4"	43	3"	250'
4"	43	4"	980'
4"	540	2"	21'
4"	540	3"	150'
4"	540	4"	580'

(This table for use with vent stacks and stack vents.)

Table A-5. Horizontal branch sizing in Zone Two

Pipe size	Maximum number of fixture-units on a horizontal branch
1¼"	1
1½"	3
2"	6
3"	20*
4"	160
6"	620

*Not more than two toilets may be connected to a single 3-inch horizontal branch. Any branch connecting with a toilet must have a minimum diameter of 3 inches.

Note: This table does not represent branches of the building drain. Other restrictions apply under battery-venting conditions.

Table A-6. Sizing a wet stack vent in Zone Two

Pipe size of stack	Fixture-unit load on stack	Maximum length of stack
2"	4	30
3"	24	50
4"	50	100
6"	100	300

Table A-7
Vent stack for wet venting in Zone Two

Wet-vented fixtures	Vent stack size requirements
1–2 bathtubs or showers	2"
3–5 bathtubs or showers	2½"
6–9 bathtubs or showers	3"
10–16 bathtubs or showers	4"

Table A-8. Vent sizing for Zone Three

Drain pipe size	Drain pipe grade per foot	Vent pipe size	Maximum developed length of vent pipe
1½"	¼"	1¼"	Unlimited
1½"	¼"	1½"	Unlimited
2"	¼"	1¼"	290'
2"	¼"	1½"	Unlimited
3"	¼"	1½"	97'
3"	¼"	2"	420'
3"	¼"	3"	Unlimited
4"	¼"	2"	98'
4"	¼"	3"	Unlimited
4"	¼"	4"	Unlimited

(This table for use with individual, branch, and circuit vents for horizontal drain pipes.)

Table A-9
Stack venting without individual vents in Zone Two

Fixtures allowed to be stack vented without individual vents*
Water closets
Basins
Bathtubs
Showers
Kitchen sinks, with or without dishwasher and garbage disposal

**Note:* Restrictions apply to this type of installation.

Table A-10
Sizing building drains in Zone Three

Pipe size	Pipe grade to the foot	Maximum number of fixture-units
2"	¼"	21
3"	¼"	42*
4"	¼"	216

**Note:* No more than two toilets may be installed on a 3-inch building drain.

Table A-11. Fixture-unit ratings in Zone Three

Bathtub	2
Shower	2
Residential toilet	4
Lavatory	1
Kitchen sink	2
Dishwasher	2
Clothes washer	3
Laundry tub	2

Table A-12. Zone One's fixture-unit requirements on trap sizes

Trap Size	Units
1¼"	One fixture
1½"	Three fixture
2"	Four fixture
3"	Six fixture
4"	Eight fixture

Table A-13. Zone Two's fixture-unit requirements on trap sizes

Trap Size	Units
1¼"	One fixture
1½"	Two fixture
2"	Three fixture
3"	Five fixture
4"	Six fixture

Table A-14. Zone Three's fixture-unit requirements on trap sizes

Trap Size	Units
1¼"	One fixture
1½"	Two fixture
2"	Three fixture
3"	Five fixture
4"	Six fixture

Table A-15. Common minimum fixture-supply sizes

Fixture	Minimum supply size
Bathtub	½"
Bidet	⅜"
Shower	½"
Toilet	⅜"
Lavatory	⅜"
Kitchen sink	½"
Dishwasher	½"
Laundry tub	½"
Hose bibb	½"

Table A-16. Recommended capacities at fixture supply outlets

Fixture	Flow rate (gpm)[a]	Flow pressure (psi)[b]
Bathtub	4	8
Bidet	2	4
Dishwasher	2.75	8
Hose bibb	5	8
Kitchen sink	2.5	8
Laundry tub	4	8
Lavatory	2	8
Shower	3	8
Water closet (two-piece)	3	8
Water closet (one-piece)	6	20

[a]gpm = gallons per minute

[b]psi = pounds per square inch

Table A-17. Minimum drain pipe pitch for Zone One

Pipes under 4" in diameter	¼" to the foot
Pipes 4" or larger in diameter	⅛" to the foot

Table A-18. Minimum drain pipe pitch for Zone Two

Pipes under 3" in diameter	¼" to the foot
Pipes 3" or larger in diameter	⅛" to the foot

Table A-19. Minimum drain pipe pitch for Zone Three

Pipes under 3" in diameter	¼" to the foot
Pipes 3" to 6" in diameter	⅛" to the foot
Pipes 8" or larger in diameter	1/16" to the foot

Table A-20. Recommended fittings for changes in direction of DWV pipes

Type of fitting	Horizontal to vertical	Vertical to horizontal	Horizontal to horizontal
Sixteenth-bend	yes	yes	yes
Eighth-bend	yes	yes	yes
Sixth-bend	yes	yes	yes
Quarter-bend	yes	no	no
Short sweep	yes	yes	no
Long sweep	yes	yes	yes
Sanitary tee	yes	no	no
Wye	yes	yes	yes
Combination wye and eighth-bend	yes	yes	yes

Table A-21. Fittings approved for horizontal to vertical changes in Zone One

45-degree wye
60-degree wye
Combination wye and eighth-bend
Sanitary tee
Sanitary tapped-tee branches

**Note:* Cross fittings, like double sanitary tees, cannot be used when they are of a short-sweep pattern; however, double sanitary tees can be used if the barrel of the tee is at least two pipe sizes larger than the largest inlet.

Table A-22. Fittings approved for vertical to horizontal changes in Zone One

45 degree branches
60-degree branches and offsets, if they are installed in a true vertical position

Table A-23. Fittings approved for horizontal changes in Zone One

45-degree wye
Combination wye and eighth-bend

**Note:* Other fittings with similar sweeps may also be approved.

Table A-24. Recommended trap sizes for Zone One

Type of fixture	Trap size
Bathtub	1½"
Shower	2"
Residential toilet	Integral
Lavatory	1¼"
Bidet	1½"
Laundry tub	1½"
Washing machine standpipe	2"
Floor drain	2"
Kitchen sink	1½"
Dishwasher	1½"
Drinking fountain	1¼"
Public toilet	Integral

Table A-25. Recommended trap sizes for Zone Two

Type of fixture	Trap size
Bathtub	1½"
Shower	2"
Residential toilet	Integral
Lavatory	1¼"
Bidet	1½"
Laundry tub	1½"
Washing machine standpipe	2"
Floor drain	2"

Table A-25. Continued

Type of fixture	Trap size
Kitchen sink	1½"
Dishwasher	1½"
Drinking fountain	1"
Public toilet	Integral

Table A-26. Recommended trap sizes for Zone Three

Type of fixture	Trap size
Bathtub	1½"
Shower	2"
Residential toilet	Integral
Lavatory	1¼"
Bidet	1¼"
Laundry tub	1½"
Washing machine standpipe	2"
Floor drain	2"
Kitchen sink	1½"
Dishwasher	1½"
Drinking fountain	1¼"
Public toilet	Integral
Urinal	2"

Table A-27 Trap-to-vent distances in Zone One

Grade on drain pipe	Size of trap arm	Maximum distance between trap and vent
¼"	1¼"	2'6"
¼"	1½"	3'6"
¼"	2"	5'
¼"	3"	6'
¼"	4" and larger	10'

Table A-28. Trap-to-vent distances in Zone Two

Grade on drain pipe	Fixture's drain size	Trap size	Maximum distance between trap and vent
¼"	1¼"	1¼"	3'6"
¼"	1½"	1¼"	5'
¼"	1½"	1½"	5'
¼"	2"	1½"	8'
¼"	2"	2"	6'
⅛"	3"	3"	10'
⅛"	4"	4"	12'

Table A-29. Trap-to-vent distances in Zone Three

Grade on drain pipe	Fixture's drain size	Trap size	Maximum distance between trap and vent
¼"	1¼"	1¼"	3'6"
¼"	1½"	1¼"	5'
¼"	1½"	1½"	5'
¼"	2"	1½"	8'
¼"	2"	2"	6'
⅛"	3"	3"	10'
⅛"	4"	4"	12'

Table A-30. Horizontal pipe-support intervals in Zone One

Type of drainage pipe	Maximum distance of supports
ABS	4'
Cast iron	At each pipe joint*
Galvanized (1" and larger)	12'
Galvanized (¾" and smaller)	10'
PVC	4'
Copper (2" and larger)	10'
Copper (1½" and smaller)	6'

*Cast-iron pipe must be supported at each joint, but supports may not be more than 10 feet apart.

Table A-31. Horizontal pipe-support intervals in Zone One

Type of vent pipe	Maximum distance of supports
ABS	4'
Cast iron	At each pipe joint*
Galvanized	12'
Copper (1½" and smaller)	6'
PVC	4'
Copper (2" and larger)	10'

*Cast-iron pipe must be supported at each joint, but supports may not be more than 10 feet apart.

Table A-32. Horizontal pipe-support intervals in Zone Two

Type of vent pipe	Maximum distance of supports
ABS	4'
Cast iron	At each pipe joint
Galvanized	12'
PVC	4'
Copper (2" and larger)	10'
Copper (1½" and smaller)	6'

Table A-33. Horizontal pipe-support intervals in Zone Three

Type of vent pipe	Maximum distance of supports
PB pipe	32"
Lead pipe	Continuous
Cast iron	5' or at each joint
Galvanized	12'
Copper tube (1¼")	6'
Copper tube (1½" and larger)	10'
ABS	4'
PVC	4'
Brass	10'
Aluminum	10'

Table A-34. Vertical pipe-support intervals in Zone One

Type of drainage pipe	Maximum distance of supports
Lead pipe	4'
Cast iron	At each story
Galvanized	At least every other story
Copper	At each story*
PVC	Not mentioned
ABS	Not mentioned

*Support intervals may not exceed 10 feet.

Note: All stacks must be supported at their bases.

Table A-35. Vertical pipe-support intervals in Zone Two

Type of drainage pipe	Maximum distance of supports
Lead pipe	4'
Cast iron	At each story*
Galvanized	At each story**
Copper (1¼" and smaller)	4'
Copper (1½" and larger)	At each story
PVC (1½" and smaller)	4'
PVC (2" and larger)	At each story
ABS (1½" and smaller)	4'
ABS (2" and larger)	At each story

*Support intervals may not exceed 15 feet.

**Support intervals may not exceed 30 feet.

Note: All stacks must be supported at their bases.

Table A-36. Vertical pipe-support intervals in Zone Three

Type of vent pipe	Maximum distance of supports
Lead pipe	4'
Cast iron	15'
Galvanized	15'
Copper tubing	10'
ABS	4'
PVC	4'
Brass	10'
Aluminum	15'

Table A-37. Support intervals for supporting water pipe in Zone One

Type of pipe	Vertical support interval	Horizontal support interval
Threaded pipe (¾" and smaller)	Every other story	10'
Threaded pipe (1" and larger)	Every other story	12'
Copper tube (1½" and smaller)	Every story, not to exceed 10'	6'
Copper tube (2" and larger)	Every story, not to exceed 10'	10'
Plastic pipe	Not mentioned	4'

Table A-38. Support intervals for supporting water pipe in Zone Two

Type of pipe	Vertical support interval	Horizontal support interval
Threaded pipe	30'	12'
Copper tube (1¼" and smaller)	4'	6'
Copper tube (1½")	Every story	6'
Copper tube (larger than 1½")	Every story	10'
Plastic pipe (2" and larger)	Every story	4'
Plastic pipe (1½" and smaller)	4'	4'

Table A-39. Materials approved for above-ground vents in Zone One

Cast-iron materials
ABS materials*
PVC materials*
Copper materials
Galvanized materials
Lead materials
Brass materials

*These materials may not be used with buildings having more than three floors above grade.

Table A-40. Materials approved for above-ground vents in Zone Two

Cast-iron materials
ABS materials
PVC materials
Copper materials
Galvanized materials
Lead materials
Aluminum materials
Borosilicate-glass materials
Brass materials

Table A-41. Materials approved for above-ground vents in Zone Three

Cast-iron materials
ABS materials
PVC materials
Copper materials
Galvanized materials
Lead materials
Aluminum materials
Brass materials

Table A-42. Materials approved for underground vents in Zone One

Cast-iron materials
ABS materials*
PVC materials*
Copper materials
Brass materials
Lead materials

*These materials may not be used with buildings having more than three floors above grade.

Table A-43. Materials approved for underground vents in Zone Two

Cast-iron materials
ABS materials
PVC materials
Copper materials
Aluminum materials
Borosilicate-glass materials

Table A-44. Materials approved for underground vents in Zone Three

Cast-iron materials
ABS materials
PVC materials
Copper materials

Table A-45. Stack-sizing for Zone Two

Pipe size	Fixture-unit discharge on stack from a branch	Total fixture-units allowed on stack
1½"	3	4
2"	6	10
3"	20*	30*
4"	160	240

*No more than two toilets may be placed on a 3-inch branch, and no more than six toilets may be connected to a 3-inch stack.

Table A-46. Stack-sizing for Zone Three

Pipe size	Fixture-unit discharge on stack from a branch	Total fixture-units allowed on stack
1½"	2	4
2"	6	10
3"	20*	48*
4"	90	240

*No more than two toilets may be placed on a 3-inch branch, and no more than six toilets may be connected to a 3-inch stack.

Table A-47. Stack-sizing tall stacks in Zone Two

Pipe size	Fixture-unit discharge on stack from a branch	Total fixture-units allowed on stack
1½"	2	8
2"	6	24
3"	16*	60*
4"	90	500

*No more than two toilets may be placed on a 3-inch branch, and no more than six toilets may be connected to a 3-inch stack.

Table A-48. Stack-sizing tall stacks in Zone Three

Pipe size	Fixture-unit discharge on stack from a branch	Total fixture-units allowed on stack
1½"	2	8
2"	6	24
3"	20*	72*
4"	90	500

*No more than two toilets may be placed on a 3-inch branch, and no more than six toilets may be connected to a 3-inch stack.

B

Sample forms

Your Company Name
Your Company Address
Your Company Phone Number

PROPOSAL

Date: ____________________________

Customer name: ____________________________

Address: ____________________________

Phone number: ____________________________

Job location: ____________________________

Description of Work

Your Company Name will supply, and or coordinate, all labor and material for the above referenced job as follows:

____

____

____

Payment Schedule

Price: ____________________________ ($__________),

Payments to be made as follows:

____

____

____

All payments shall be made in full, upon presentation of each completed invoice. If payment is not made according to the terms above, Your Company Name will have the following rights and remedies. Your Company Name may charge a monthly service charge of one-and-one-half percent (1.5%), eighteen percent (18%) per year, from the first day default is made. Your Company Name may lien the property where the work has been done. Your Company Name may use all legal methods in the collection of monies owed to it. Your Company Name may seek compensation, at the rate of $__________ per hour, for attempts made to collect unpaid monies.

Page 1 of 2 initials __________

B-1 *Proposal*

Your Company Name may seek payment for legal fees and other costs of collection, to the full extent the law allows.

If the job is not ready for the service or materials requested, as scheduled, and the delay is not due to Your Company Name's actions, Your Company Name may charge the customer for lost time. This charge will be at a rate of $_______ per hour, per man, including travel time.

If you have any questions or don't understand this proposal, seek professional advice. Upon acceptance, this proposal becomes a binding contract between both parties.

Respectfully submitted,

Your name and title
Owner

Acceptance

We the undersigned do hereby agree to, and accept, all the terms and conditions of this proposal. We fully understand the terms and conditions, and hereby consent to enter into this contract.

Your Company Name	Customer
By ______________________	______________________
Title______________________	Date______________________
Date______________________	

Proposal expires in 30 days, if not accepted by all parties.

B-1 *Continued*

Renaissance Remodeling
357 Paris Lane
Wilton, Ohio 55555
(102) 555-5555

REMODELING CONTRACT

This agreement, made this _____th day of ____________________, 19____, shall set forth the whole agreement, in its entirety, between Contractor and Customer.

Contractor: Renaissance Remodeling, referred to herein as Contractor.

Customer: ____, referred to herein as Customer.

Job name: ____

Job location: ____

The Customer and Contractor agree to the following:

Scope of Work

Contractor shall perform all work as described below and provide all material to complete the work described below: All work is to be completed by Contractor in accordance with the attached plans and specifications. All material is to be supplied by Contractor in accordance with attached plans and specifications. Said attached plans and specifications have been acknowledged and signed by Contractor and Customer.

A brief outline of the work is as follows, and all work referenced in the attached plans and specifications will be completed to the Customer's reasonable satisfaction. The following is only a basic outline of the overall work to be performed:

____

____

____

____

____

____

____

____

____

(Page 1 of 3 initials_______________)

B-2 *Remodeling contract*

Commencement and Completion Schedule

The work described above shall be started within three days of verbal notice from Customer; the projected start date is ____________. The Contractor shall complete the above work in a professional and expedient manner, by no later than _____ days from the start date. Time is of the essence regarding this contract. No extension of time will be valid, without the Customer's written consent. If Contractor does not complete the work in the time allowed, and if the lack of completion is not caused by the Customer, the Contractor will be charged _______, per day, for every day work is not finished beyond the completion date. This charge will be deducted from any payments due to the Contractor for work performed.

Contract Sum

The Customer shall pay the Contractor for the performance of completed work, subject to additions and deductions, as authorized by this agreement or attached addendum. The contract sum is ____, ($______________).

Progress Payments

The Customer shall pay the Contractor installments as detailed below, once an acceptable insurance certificate has been filed by the Contractor, with the Customer:

Customer will pay Contractor a deposit of ____,
($____________), when work is started.
Customer will pay ____,
($___________), when all rough-in work is complete.
Customer will pay ____,
($_______________) when work is __________ percent complete.
Customer will pay ____,
($____________) when all work is complete and accepted.

All payments are subject to a site inspection and approval of work by the Customer. Before final payment, the Contractor, if required, shall submit satisfactory evidence to the Customer, that all expenses related to this work have been paid and no lien risk exists on the subject property.

Working Conditions

Working hours will be ___ A.M. through ___ P.M., Monday through Friday. Contractor is required to clean work debris from the job site on a daily basis and to leave the site in a clean and neat condition. Contractor shall be responsible for removal and disposal of all debris related to their job description.

(Page 2 of 3 initials____________)

B-2 *Continued*

Contract Assignment

Contractor shall not assign this contract or further subcontract the whole of this subcontract without the written consent of the Customer.

Laws, Permits, Fees, and Notices

Contractor is responsible for all required laws, permits, fees, or notices required to perform the work stated herein.

Work of Others

Contractor shall be responsible for any damage caused to existing conditions. This shall include work performed on the project by other contractors. If the Contractor damages existing conditions or work performed by other contractors, said Contractor shall be responsible for the repair of said damages. These repairs may be made by the Contractor responsible for the damages or another contractor, at the sole discretion of Customer.

The damaging Contractor shall have the opportunity to quote a price for the repairs. The Customer is under no obligation to engage the damaging Contractor to make the repairs. If a different contractor repairs the damage, the Contractor causing the damage may be back-charged for the cost of the repairs. These charges may be deducted from any monies owed to the damaging Contractor.

If no money is owed to the damaging Contractor, said Contractor shall pay the invoiced amount within _____ business days. If prompt payment is not made, the Customer may exercise all legal means to collect the requested monies. The damaging Contractor shall have no rights to lien the Customer's property for money retained to cover the repair of damages caused by the Contractor. The Customer may have the repairs made to his satisfaction.

Warranty

Contractor warrants to the Customer all work and materials, for one year from the final day of work performed.

Indemnification

To the fullest extent allowed by law, the Contractor shall indemnify and hold harmless the Customer and all of their agents and employees from and against all claims, damages, losses and expenses.

This Agreement entered into on __________________________, 19_____ shall constitute the whole agreement between Customer and Contractor.

______________________________ ______________________________

Customer Date Contractor Date

____

Customer Date

B-2 *Continued*

Commencement and Completion Schedule

The work described above shall be started within three (3) days of verbal notice from the customer, the projected start date is ______________________. The subcontractor shall complete the above work in a professional and expedient manner by no later than twenty (20) days from the start date.
Time is of the essence in this subcontract. No extension of time will be valid without the general contractor's written consent. If subcontractor does not complete the work in the time allowed and if the lack of completion is not caused by the general contractor, the subcontractor will be charged one-hundred dollars ($100.00) for every day work is not finished after the completion date. This charge will be deducted from any payments due to the subcontractor for work performed.

B-3 *Sample completion clause*

Subcontractor Liability for Damages

Subcontractor shall be responsible for any damage caused to existing conditions. This shall include new work performed on the project by other contractors. If the subcontractor damages existing conditions or work performed by other contractors, said subcontractor shall be responsible for the repair of said damages. These repairs may be made by the subcontractor responsible for the damages or another contractor, at the discretion of the general contractor.

If a different contractor repairs the damage, the subcontractor causing the damage may be back-charged for the cost of the repairs. These charges may be deducted from any monies owed to the damaging subcontractor, by the general contractor. The choice for a contractor to repair the damages shall be at the sole discretion of the general contractor.

If no money is owed to the damaging subcontractor, said contractor shall pay the invoiced amount, to the general contractor, within seven (7) business days. If prompt payment is not made,
the general contractor may exercise all legal means to collect the requested monies.

The damaging subcontractor shall have no rights to lien the property where work is done for money retained to cover the repair of damages caused by the subcontractor. The general contractor may have the repairs made to his satisfaction.

The damaging subcontractor shall have the opportunity to quote a price for the repairs. The general contractor is under no obligation to engage the damaging subcontractor to make the repairs.

B-4 *Sample damage clause for contract*

Certificate of Completion and Acceptance

Contractor: ____

Customer: ____

Job name: ____

Job location: ____

Job Description: ____

Date of completion: ____

Date of final inspection by customer: ____

Date of code compliance inspection & approval: ____

Defects found in material or workmanship: ____

Acknowledgment

Customer acknowledges the completion of all contracted work and accepts all workmanship and materials as being satisfactory. Upon signing this certificate, the customer releases the contractor from any responsibility for additional work, except warranty work. Warranty work will be performed for a period of one year from the date of completion. Warranty work will include the repair of any material or workmanship defects occurring between now and the end of the warranty period. All existing workmanship and materials are acceptable to the customer and payment will be made, in full, according to the payment schedule in the contract, between the two parties.

B-5 *Certificate of completion and acceptance*

Certificate of Subcontractor Completion Acceptance

Contractor: ____

Subcontractor: ____

Job name: ____

Job location: ____

Job description: ____

____

____

____

____

____

Date of completion: ____

Date of final inspection by contractor: ______________________________

Date of code compliance inspection & approval: ________________________

Defects found in material or workmanship: ____________________________

____

____

____

____

Acknowledgment

Contractor acknowledges the completion of all contracted work and accepts all workmanship and materials as being satisfactory. Upon signing this certificate, the contractor releases the subcontractor from any responsibility for additional work, except warranty work. Warranty work will be performed for a period of one year from the date of completion. Warranty work will include the repair of any material or workmanship defects occurring between now and the end of the warranty period. All existing workmanship and materials are acceptable to the contractor and payment will be made, in full, according to the payment schedule in the contract, between the two parties.

______________________________ ______________________________

Contractor Date Subcontractor Date

B-6 *Certificate of subcontractor completion acceptance*

Addendum

This addendum is an integral part of the contract dated ______________________, between the Contractor, ____, and the Customer(s), ____, for the work being done on real estate commonly known as __________________________. The undersigned parties hereby agree to the following:

____

____

____

____

____

____

____

____

____

____

____

____

____

____

____

The above constitutes the only additions to the above-mentioned contract, no verbal agreements or other changes shall be valid unless made in writing and signed by all parties.

______________________________	______________________________
Contractor Date	Customer Date
	______________________________
	Customer Date

B-7 *Addendum*

Change Order

This change order is an integral part of the contract dated______________________, between the customer, __________________________________ , and the contractor,_____________________, for the work to be performed. The job location is ____. The following changes are the only changes to be made. These changes shall now become a part of the original contract and may not be altered again without written authorization from all parties.

Changes to be as follows:

____

____

____

____

____

____

These changes will increase/decrease the original contract amount. Payment for theses changes will be made as follows:____________________________________. The amount of change in the contract price will be ______________________________ ($ ________). The new total contract price shall be __________________________ ($ ________).

The undersigned parties hereby agree that these are the only changes to be made to the original contract. No verbal agreements will be valid. No further alterations will be allowed without additional written authorization, signed by all parties. This change order constitutes the entire agreement between the parties to alter the original contract.

________________________ Customer	________________________ Contractor
________________________ Date	________________________ Date
________________________ Customer	
________________________ Date	

B-8 *Change order*

Green Tree Lawn Care
987 Willow Road
Wilson, Maine 55555
(101) 555-5555

WORK ESTIMATE

Date: ______________________

Customer name: ____

Address: ____

Phone number: ____

Description of Work

Green Tree Lawn Care will supply all labor and material for the following work:

____

____

____

____

____

____

____

____

Payment for Work as Follows

Estimated price: ________________________________, payable as follows

____

____

If you have any questions, please don't hesitate to call. Upon acceptance, a formal contract will be issued.

Respectfully submitted,

J. B. Williams
Owner

B-9 *Work estimate*

Your Company Name
Your Company Address
Your Company Phone Number

Quote

This agreement, made this ____________ day of ___________ , 19______, shall set forth the whole agreement, in its entirety, by and between Your Company Name, herein called Contractor and___________, herein called Owners.

Job name: ____

Job location: ____

The Contractor and Owners agree to the following:

Contractor shall perform all work as described below and provide all material to complete the work described below. Contractor shall supply all labor and material to complete the work according to the attached plans and specifications. The work shall include the following:

____

____

____

Schedule

The work described above shall begin within three days of notice from Owner, with an estimated start date of _______________. The Contractor shall complete the above work in a professional and expedient manner within ___ days from the start date.

Payment Schedule

Payments shall be made as follows:

____

____

This agreement, entered into on ____________, shall constitute the whole between Contractor and Owner.

Contractor	Date	Owner	Date
		Owner	Date

B-10 *Quote*

Repair Voucher

Date______________________

Time______________________

Received of______________________

Address______________________

Phone number______________________

Item to be repaired______________________

Serial number______________________

Make______________________

Model______________________

Nature of problem______________________

Item accepted by______________________

B-11 *Repair voucher*

Subcontract Agreement

This agreement, made this ____th day of ____________, 19___, shall set forth the whole agreement, in its entirety, between Contractor and Subcontractor.
Contractor: ____, referred to herein as Contractor.

Job location: ____

Subcontractor: ____, referred to herein as Subcontractor.
The Contractor and Subcontractor agree to the following:

Scope of Work

Subcontractor shall perform all work as described below and provide all material to complete the work described below.

Subcontractor shall supply all labor and material to complete the work according to the attached plans and specifications. These attached plans and specifications have been initialed and signed by all parties. The work shall include, but is not limited to, the following:

____

____

____

____

____

____

____

____

Commencement and Completion Schedule

The work described above shall be started within three days of verbal notice from Contractor, the projected start date is ______________________. The Subcontractor shall complete the above work in a professional and expedient manner by no later than _____ days from the start date. Time is of the essence in this contract. No extension of time will be valid without the Contractor's written consent. If Subcontractor does not complete the work in the time allowed, and if the lack of completion is not caused by the Contractor, the Subcontractor will be charged fifty dollars ($50.00) per day, for every day work extends beyond the completion date. This charge will be deducted from any payments due to the Subcontractor for work performed.

Page 1 of 3 initials___

B-12 *Subcontract agreement*

Contract Sum

The Contractor shall pay the Subcontractor for the performance of completed work subject to additions and deductions as authorized by this agreement or attached addendum. The contract sum is ______________________________($____________).

Progress Payments

The Contractor shall pay the Subcontractor installments as detailed below, once an acceptable insurance certificate has been filed by the Subcontractor with the Contractor.
Contractor shall pay the Subcontractor as described:

____

____

____

____

____

____

____

____

____

____

____

____

____

____

____

____

____

All payments are subject to a site inspection and approval of work by the Contractor. Before final payment, the Subcontractor shall submit satisfactory evidence to the Contractor that no lien risk exists on the subject property.

Page 2 of 3 initials___

B-12 *Continued*

Working Conditions

Working hours will be 8:00 A.M. through 4:30 P.M., Monday through Friday. Subcontractor is required to clean his work debris from the job site on a daily basis and leave the site in a clean and neat condition. Subcontractor shall be responsible for removal and disposal of all debris related to his job description.

Contract Assignment

Subcontractor shall not assign this contract or further subcontract the whole of this subcontract, without the written consent of the Contractor.

Laws, Permits, Fees, and Notices

Subcontractor shall be responsible for all required laws, permits, fees, or notices, required to perform the work stated herein.

Work of Others

Subcontractor shall be responsible for any damage caused to existing conditions or other contractor's work. This damage will be repaired, and the Subcontractor charged for the expense and supervision of this work. The Subcontractor shall have the opportunity to quote a price for said repairs, but the Contractor is under no obligation to engage the Subcontractor to make said repairs. If a different subcontractor repairs the damage, the Subcontractor may be back-charged for the cost of the repairs. Any repair costs will be deducted from any payments due to the Subcontractor. If no payments are due the Subcontractor, the Subcontractor shall pay the invoiced amount within 10 days.

Warranty

Subcontractor warrants to the Contractor, all work and materials for one year from the final day of work performed.

Indemnification

To the fullest extent allowed by law, the Subcontractor shall indemnify and hold harmless the Owner, the Contractor, and all of their agents and employees from and against all claims, damages, losses and expenses.

This agreement, entered into on ____________________________, 19______, shall constitute the whole agreement between Contractor and Subcontractor.

______________________________		______________________________	
Contractor	Date	Subcontractor	Date

B-12 *Continued*

Code Violation Notification

Contractor: ____________________

Contractor's address: ____________________

City/state/zip: ____________________

Phone number: ____________________

Job location: ____________________

Date: ____________________

Type of work: ____________________

Subcontractor: ____________________

Address: ____________________

Official Notification of Code Violations

On March 22, 1993, I was notified by the local code enforcement officer of code violations in the work performed by your company. The violations must be corrected within two business days, as per our contract dated March 1, 1993. Please contact the codes officer for a detailed explanation of the violations and required corrections. If the violations are not corrected within the allotted time, you may be penalized, as per our contract, for your actions in delaying the completion of this project. Thank you for your prompt attention to this matter.

____

General Contractor Date

B-13 *Code violation notification*

Letter of Engagement

Client ____

Street ____

City/State/Zip ____

Work phone ______________________ Home phone ______________________

Services requested____

____

____

____

____

____

Fee for services described above $____

Payment to be made as follows:

____

____

____

By signing this letter of engagement, you indicate your understanding that this engagement letter constitutes a contractual agreement between us for the services set forth. This engagement does not include any services not specifically stated in this letter. Additional services, which you may request, will be subject to separate arrangements, to be set forth in writing.

A representative of ______________________ has advised us that we should seek legal counsel prior to using information or material received from ______________________.

We the undersigned hereby release ______________________, its employees, officers, shareholders, and representatives from any liability. We understand that we shall have no rights, claims, or recourse and waive any claims or rights we may have against ______________________, its employees, officers, shareholders, and representatives. We further understand that we will pay all costs of collection of any amount due hereunder including reasonable attorney fees.

______________________ ______________________

Client Date Client Date

____

Company Representative Date

B-14 *Letter of engagement*

Short-Form Lien Waiver

Customer name: ______________________________

Customer address: ______________________________

Customer city/state/zip: ______________________________

Customer phone number: ______________________________

Job location: ______________________________

Date: ______________________________

Type of work: ______________________________

Contractor: ______________________________

Contractor address: ______________________________

Subcontractor: ______________________________

Subcontractor address: ______________________________

Description of work completed to date: ______________________________

____

____

____

Payments received to date: ______________________________

Payment received on this date: ______________________________

Total amount paid, including this payment: ______________________________

The contractor/subcontractor signing below acknowledges receipt of all payments stated above. These payments are in compliance with the written contract between the parties above. The contractor/subcontractor signing below hereby states payment for all work done to this date has been paid in full.

The contractor/subcontractor signing below releases and relinquishes any and all rights available to place a mechanic or materialman lien against the subject property for the above described work. All parties agree that all work performed to date has been paid for in full and in compliance with their written contract.

The undersigned contractor/subcontractor releases the general contractor/customer from any liability for non-payment of material or services extended through this date. The undersigned contractor/subcontractor has read this entire agreement and understands the agreement.

____

Contractor/Subcontractor Date

B-15 *Short form lien waiver*

Long-Form Lien Waiver

Customer name: ____________________

Customer address: ____________________

Customer city/state/zip: ____________________

Customer phone number: ____________________

Job location: ____________________

Date: ____________________

Type of work: ____________________

The vendor acknowledges receipt of all payments stated below. These payments are in compliance with the written contract between the vendor and the customer. The vendor hereby states that payment for all work done to this date has been paid in full.

The vendor releases and relinquishes any and all rights available to said vendor to place a mechanic or materialman lien against the subject property for the described work. Both parties agree that all work performed to date has been paid for, in full and in compliance with their written contract.

The undersigned vendor releases the customer and the customer's property from any liability for non-payment of material or services extended through this date. The undersigned contractor has read this entire agreement and understands the agreement.

Vendor Name	Signature of Co. Rep.	Signature Date	Service Performed	Date Paid	Amount Paid
Plumber (Rough-in)					
Plumber (Final)					
Electrician (Rough-in)					
Electrician (Final)					
Supplier (Framing lumber)					

*This list should include all contractors and suppliers. All vendors are listed on the same lien waiver, and sign next to their trade name for each service rendered, at the time of payment.

B-16 *Long form lien waiver*

Your Company Name
Your Company Address

Dear Sir:

I am soliciting bids for the work listed below, and I would like to offer you the opportunity to participate in the bidding. If you are interested in giving quoted prices on material for this job, please let me hear from you, at the above address.

The job will be started in _______ weeks. Financing has been arranged and the job will be started on schedule. Your quote, if you choose to enter one, must be received no later than ________________________.

The proposed work is as follows:

____

____

____

Plans and specifications for the work are available upon request.

Thank you for your time and consideration in this request.

Sincerely,

Your name and title

B-17 *Form letter for soliciting material quotes*

Material Order Log

Supplier: ______________________________

Date order was placed: ______________________________

Time order was placed: ______________________________

Name of person taking order: ______________________________

Promised delivery date: ______________________________

Order number: ______________________________

Quoted price: ______________________________

Date of follow-up call: ______________________________

Manager's name: ______________________________

Time of call to manager: ______________________________

Manager confirmed delivery date: ______________________________

Manager confirmed price: ______________________________

Notes and Comments

____

____

____

____

____

____

____

____

____

____

____

____

B-18 *Material order log*

Customer Reference Report

Date:
Job Name:
Job Address:

Quality of workmanship	Poor	Good	Great
Quality of materials	Poor	Good	Great
Dependability	Poor	Good	Great
Overall satisfaction	Poor	Good	Great
Would recommend to others	No	Maybe	Yes

Comments

____

____

____

____

Customer signature

B-19 *Job performance report*

SOLICITATION MAILING LIST APPLICATION

1. TYPE OF APPLICATION ☐ INITIAL ☐ REVISION | 2. DATE | FORM APPROVED OMB NO. 3090-0009

NOTE—Please complete all items on this form. Insert N/A in items not applicable. See reverse for Instructions.

3. NAME AND ADDRESS OF FEDERAL AGENCY TO WHICH FORM IS SUBMITTED *(Include ZIP code)*

4. NAME AND ADDRESS OF APPLICANT *(Include county and ZIP code)*

5. TYPE OF ORGANIZATION *(Check one)*

☐ INDIVIDUAL ☐ NON-PROFIT ORGANIZATION

☐ PARTNERSHIP ☐ CORPORATION, INCORPORATED UNDER THE LAWS OF THE STATE OF:

6. ADDRESS TO WHICH SOLICITATIONS ARE TO BE MAILED *(If different than Item 4)*

7. NAMES OF OFFICERS, OWNERS, OR PARTNERS

A. PRESIDENT | B. VICE PRESIDENT | C. SECRETARY

D. TREASURER | E. OWNERS OR PARTNERS

8. AFFILIATES OF APPLICANT *(Names, locations and nature of affiliation. See definition on reverse.)*

9. PERSONS AUTHORIZED TO SIGN OFFERS AND CONTRACTS IN YOUR NAME *(Indicate if agent)*

NAME	OFFICIAL CAPACITY	TELE. NO. *(Include area code)*

10. IDENTIFY EQUIPMENT, SUPPLIES, AND/OR SERVICES ON WHICH YOU DESIRE TO MAKE AN OFFER *(See attached Federal agency's supplemental listing and instructions, if any)*

11A. SIZE OF BUSINESS *(See definitions on reverse)*

☐ SMALL BUSINESS *(If checked, complete items 11B and 11C)* ☐ OTHER THAN SMALL BUSINESS

11B. AVERAGE NUMBER OF EMPLOYEES *(Including affiliates)* FOR FOUR PRECEDING CALENDAR QUARTERS

11C. AVERAGE ANNUAL SALES OR RECEIPTS FOR PRECEDING THREE FISCAL YEARS $

12. TYPE OF OWNERSHIP *(See definitions on reverse) (Not applicable for other than small businesses)*

☐ DISADVANTAGED BUSINESS ☐ WOMAN-OWNED BUSINESS

13. TYPE OF BUSINESS *(See definitions on reverse)*

☐ MANUFACTURER OR PRODUCER ☐ REGULAR DEALER *(Type 1)* ☐ CONSTRUCTION CONCERN ☐ SURPLUS DEALER

☐ SERVICE ESTABLISHMENT ☐ REGULAR DEALER *(Type 2)* ☐ RESEARCH AND DEVELOPMENT

14. DUNS NO. *(If available)* | 15. HOW LONG IN PRESENT BUSINESS?

16. FLOOR SPACE *(Square feet)*: A. MANUFACTURING | B. WAREHOUSE

17. NET WORTH: A. DATE | B. AMOUNT $

18. SECURITY CLEARANCE *(If applicable, check highest clearance authorized)*

FOR	TOP SECRET	SECRET	CONFIDENTIAL	C. NAMES OF AGENCIES WHICH GRANTED SECURITY CLEARANCES *(Include dates)*
A. KEY PERSONNEL				
B. PLANT ONLY				

CERTIFICATION — I certify that information supplied herein *(Including all pages attached)* is correct and that neither the applicant nor any person *(Or concern)* in any connection with the applicant as a principal or officer, so far as is known, is now debarred or otherwise declared ineligible by any agency of the Federal Government from making offers for furnishing materials, supplies, or services to the Government or any agency thereof.

19. NAME AND TITLE OF PERSON AUTHORIZED TO SIGN *(Type or print)* | 20. SIGNATURE | 21. DATE SIGNED

NSN 7540-01-152-8086
PREVIOUS EDITIONS UNUSABLE

129-106

STANDARD FORM 129 (REV. 10-83)
Prescribed by GSA
FAR (48 CFR) 53.214(c)

B-20 *Form used to get on government bid lists.* U.S. Government Printing Office

Glossary

accessible Accessible, as it relates to plumbing, refers to a means of access. For example, a tub waste is considered accessible when there is an access panel that can be opened or removed to gain access to the tub waste.

air-break An air-break in the drainage system refers to an indirect waste procedure. The indirect waste enters a receptor through open air.

air-gap (drainage) A drainage air-gap refers to the vertical distance that waste travels through open air between the waste pipe and the indirect-waste receptor.

air-gap (potable water) A potable water air-gap is the vertical distance water travels through open air between the water source and the flood level rim of its receptor or fixture.

air-gap (the device) An air-gap can mean a device used to connect the drainage of a dishwasher with the sanitary drainage system.

anti-siphon Anti-siphon simply means that a device cannot be made to form a siphonic action.

area drain An area drain is a drain used to receive surface water, which can come from grounds, parking, or other areas.

aspirator An aspirator is a device used to create a vacuum, like in a suction system for medical offices.

back-flow Back-flow is the backwards flow of water or other liquids in the drainage or water system.

back-flow preventer A device used to prevent back-flow.

back-siphonage Back-siphonage is essentially the same as back-flow; it is the reverse flow of water or liquids in a pipe, caused by siphonic action.

back-water valve A back-water valve is a device installed on drainage systems to prevent a back-flow from the main sewer into the building where the valve is installed.

ballcock A ballcock is an automatic fill device. Ballcocks are most commonly found in toilet tanks. They supply a regulated amount of water, on demand, and then cut off when the water level reaches a desired height.

branch A branch is a part of the plumbing system that is not a riser, main, or stack.

branch interval A branch interval is a means of measurement for vertical waste or soil stacks. A branch interval is equal to each floor level or story in a building, but they are always at least 8 feet in height.

branch vent Branch vents are vents that connect either individual or multiple vents with a vent stack or stack vent.

building drain The building drain is the primary drainage pipe inside a building.

building sewer A building sewer is the pipe extending from the building drain to the main sewer. Building sewers usually begin between 2 and 5 feet outside of a building's foundation.

building trap A building trap is a trap installed on the building drain to prevent air from circulating between the building drain and sewer.

cistern A cistern is a covered container used to store water, normally nonpotable water.

code The code is a set of regulations that govern the installation of plumbing.

code officer A code officer is an individual responsible for enforcing the code.

combination waste and vent system A combination waste and vent system is a plumbing system where few vertical vents are used. In this type of system, the drainage pipes are often oversized to allow air to circulate in the system.

critical level Critical level refers to the point where a vacuum breaker may be submerged before back-flow can occur.

cross-connection A cross-connection is a connection or situation that may allow the contents of separate pipes to commingle.

developed length The developed length is a method of measurement that is based on the total liner footage of all pipe and fittings.

drain A drain is a pipe that conveys waste water or water-borne wastes.

drainage system A drainage system consists of all plumbing that carries sewage, rain water, and other liquid wastes to a disposal site. A drainage system does not include public sewers or sewage treatment and disposal sites.

existing work Existing work is work that was installed prior to the adoption of current code requirements.

fixture supply The water supply between a fixture and a water distribution pipe.

fixture unit A fixture unit is a unit of measure assigned to fixtures for both drainage and water and used in pipe sizing.

flood level rim The flood level rim is the point of a fixture where its contents will spill over the rim.

grade A grade is the downward fall of a pipe.

groundwork Groundwork consists of the plumbing installed below a finished grade or floor.

hot water Hot water is considered to be that which has a temperature of 110 degrees F or more.

house trap A house trap is the same as a building trap.

interceptor An interceptor is a device used to separate and retain substances not wanted in the sanitary drainage system.

leader A leader is an exterior pipe used to carry storm water from a roof or gutter.

local vent A local vent is a vertical vent used with clinical sinks to transport vapors and odors to the outside air.

main A main is any primary pipe used for water service, distribution, or drainage.

main sewer The public sewer.

main vent The main vent is the primary vent for a plumbing system.

nonpotable water Nonpotable water is water not safe for human consumption.

offset An offset is a change in direction.

open air The air outside of a building.

pitch See grade.

plumbing inspector See code officer.

potable water Potable water is water safe for human consumption in drinking, cooking, and domestic uses.

private sewage disposal system A private sewage disposal system is one that serves a private party. A septic system is an example of a private sewage disposal system.

private water supply A private water supply is a supply that serves a private party, such as a well.

readily accessible Having direct and immediate access to an object means that something is readily accessible. If an access panel must be removed before the object can be accessed, the object is not readily accessible.

rim The rim is the open edge of a fixture.

riser A riser is a vertical pipe in the water distribution system that runs vertically for at least one story.

rough-in A rough-in is the installation of plumbing in areas that will be concealed once the building is completed.

sewage Liquid waste that contains animal or vegetable matter is called sewage. The matter may be contained in either solution or suspension and may contain chemicals in solution.

slope See grade.

stack A stack is a vertical pipe in the drainage system. A stack may be a vent, soil pipe, or waste pipe.

stack vent A stack vent is a vertical pipe that extends above the highest drainage point to vent the drainage system.

storm water Storm water is another name for rain water.

sump vent A sump vent is a vertical vent that rises to vent a sump, such as in a sewer ejector sump.

tempered water Water that is tempered to maintain a temperature between 85 degrees F and 110 degrees F.

underground plumbing See groundwork.

vacuum A vacuum is pressure that is less than that produced by the atmosphere.

vacuum breaker A vacuum breaker is a device used to prevent a vacuum.

vent stack A vent stack is a vertical pipe in the drainage system that acts only as a vent and does not receive the discharge of plumbing fixtures.

waste A discharge from fixtures and equipment that does not contain fecal matter is called waste.

water distribution system The water distribution system is the collection of water pipes within a building that deliver water to fixtures and equipment.

water main The water main is the public water service.

water hammer arrester A water hammer arrester is a device that defeats water hammer by absorbing pressure surges.

water service The pipe delivering water from a water source to the water distribution system of a building is called the water service.

well A well is a water source from a hole in the ground.

wet vent A wet vent is a pipe that receives the drainage from plumbing and does double duty by venting part of the plumbing system.

yoke vent A yoke vent is an upward connection from a soil or waste stack to a vent stack. Yoke vents are normally made with wye fittings to prevent pressure changes in the stack.

Index

Illustrations are in **boldface.**

About the author

R. Dodge Woodson has nearly 20 years' experience as a homebuilder, contractor, master plumber, and real estate broker. He is also the author of many books from McGraw-Hill, including *Home Plumbing Illustrated*, *Plumbing Contractor: Start and Run a Money-Making Business*, *Troubleshooting & Repairing Heat Pumps*, *National Plumbing Codes Handbook*, *Professional Modeler's Manual: Save Time, Avoid Mistakes, Increase Profits*, and *Master Plumber's Licensing Exam Guide*.

Other Best seller's of Related Interest

Builder's Guide to Foundations & Floor Framing

Dan Ramsey

Comprehensive and knowledgeable guide that takes the reader from foundation basics through floor framing and shows how to turn an improving market into profits.

$44.00 Hardcover, ISBN 0-07-051814-9

Builder's Guide to Barriers: Doors, Windows, and Trim

Dan Ramsey

Comprehensive and knowledgeable guide that takes the reader from barrier and wall framing through to installation and finish trim.

$44.00 Hardcover, ISBN 0-07-050833-X

Builder's Guide to Change-of-Use Properties

R. Dodge Woodson

Hands-on guide that covers all aspects of remodeling single-family homes into lucrative multi-family rental properties.

$44.00 Hardcover, ISBN 0-07-071789-3

Builder's Guide to Decks

Leon A. Frechette

Offers a wealth of practical, field-tested advice on how to evaluate and bid jobs; effectively communicate with customers; understand environmental issues; work with new tools, products, and alternative materials; and perform jobs in the proper sequence.

$44.00 Hardcover, ISBN 0-07-015749-9

Builder's Guide to Wells and Septic Systems

R. Dodge Woodson

Chapters cover virtually every aspect of wells and septic systems, including on-site evaluations; site limitations; bidding; soil studies, septic designs, and code related issues; drilled and dug wells, gravel and pipe, chamber-type, and gravity septic systems; pump stations; common problems with well installation; and remedies for poor septic situations.

$44.00 Hardcover, ISBN 0-07-071782-6

The Plumber's Troubleshooting Guide

R. Dodge Woodson

This guide explains the time- and money-saving techniques used to troubleshoot all types of plumbing systems, which are explained in great detail.

$42.50 Hardcover, ISBN 0-07-071777-X